W9-BEO-888

**FOURTH
EDITION**

Effective Helping
Interviewing and Counseling Techniques

FOURTH EDITION

Effective Helping
Interviewing and Counseling Techniques

Barbara F. Okun
Northeastern University

Brooks/Cole Publishing Company
Pacific Grove, California

Brooks/Cole Publishing Company

A Division of Wadsworth, Inc.

© 1992, 1987, 1982, 1976 by Wadsworth, Inc., Belmont, California 94002. All rights reserved. No part of this book may be reproduced, stored in a retrieval system, or transcribed, in any form or by any means—electronic, mechanical, photocopying, recording, or otherwise—without the prior written permission of the publisher, Brooks/Cole Publishing Company, Pacific Grove, California 93950, a division of Wadsworth, Inc.

Printed in the United States of America

10   9   8   7   6   5   4

Library of Congress Cataloging-in-Publication Data

Okun, Barbara F.
   Effective helping : interviewing and counseling techniques /
Barbara F. Okun. — 4th ed.
      p.   cm.
   Includes bibliographical references and index.
   ISBN 0-534-14544-2
   1. Counseling. 2. Helping behavior.   I. Title.
BF637.C6054   1991
158'.3—dc20   91-23231
                                                                        CIP

Sponsoring Editor: *Claire Verduin*
Editorial Assistant: *Gay C. Bond*
Production Editor: *Penelope Sky*
Manuscript Editor: *Lorraine Anderson*
Interior and Cover Design: *Flora Pomeroy*
Art Coordinator: *Cloyce Wall*
Interior Illustration: *Cloyce Wall*
Index: *Do Mi Stauber*
Typesetting: *OmegaType Typography*
Cover Printing: *Phoenix Color Corporation*
Printing and Binding: *Arcata Graphics*

In Memoriam
Catherine A. Brenner
Katherine M. Newman

 # Preface

Effective Helping: Interviewing and Counseling Techniques provides an overview of fundamental counseling theory, as well as an introduction to interviewing and counseling that is both conceptual and experiential. Intended for undergraduate, graduate, and professional students, instructors, supervisors and trainers, and managers and administrators at all levels, this book offers the knowledge and skills that apply to professional and personal relationships.

Much has changed in our society and in the helping field since the first edition was published in the mid-1970s. The United States is now multicultural, and we have experienced population shifts and growth, overdevelopment, and homelessness and lack of health care throughout the country. We no longer have consensus in family and social values, and our confidence in such major systems as health, education, justice, and government has eroded. The human service professions and paraprofessions, which expanded during the 1970s and 1980s, are now being consolidated and even eliminated as we struggle desperately with the results of governmental deficits. Cost effectiveness has become more important than helpee welfare, although there is more need than ever for effective helping relationships as limitations on life style choices and economic opportunities affect our society. Our ingrained attitudes and historically-based expectations conflict with reality.

My personal and professional development during the past fifteen years have also influenced my views and practice. I've come to appreciate the variety of helping approaches while maintaining my belief in establishing an empathic foundation through responsive listening. New research provides new perspectives, and I've become more attuned to issues involving gender, cultural diversity, and the pressures inherent in our social system.

New to This Edition

The fourth edition of *Effective Helping* reflects many social and personal changes. While retaining the successful format and content of the previous editions, I have added several important topics. In Chapter One I have expanded upon the changes in the sociocultural and human service professions. In Chapter Five I have enlarged the sections on feminist and systems theory and updated other theories where appropriate; for example, Ivey's work on cognitive development is now in the section on eclectic theory. Chapter Eight includes more material on disaster theory. In Chapter Nine I highlight current ethical issues, such as dual role relationships, misrepresentation, conflict of interest, and financial constraints on the availability of helping services. Throughout the text I have incorporated new examples and cases, some referring to the 1991 Persian Gulf war and sociocultural issues.

The Human Relations Counseling Model

The human relations counseling model brings together the skills, helping stages, and issues involved in the helping process. Communication skills include the ability to hear and understand verbal and nonverbal messages, and to listen responsively to both kinds of messages within an empathic context. These skills enable helpers to progress through the two stages of helping: the relationship stage, during which rapport and trust develop and problems are clarified, and the strategy stage, which involves selecting and applying appropriate helping methods. Issues that may arise during the counseling process include values clarification, ethics, sexism, ethnicity, and other social and professional topics that can have either a positive or a negative impact on a helping relationship. Knowledge of the major aspects of the human relations counseling model is crucial to a helper's effectiveness in all types of settings and with all kinds of people.

In Chapters One through Four I present the three-dimensional human relations counseling model and explore the nature of the helping relationship, emphasizing both extensive theory and actual practice. In Chapters Five through Seven I describe the major theories, approaches, and strategies of helping; in Chapter Eight these strategies are applied to crisis and disaster situations. In Chapter Nine I discuss helpers' personal values and the ethical and professional issues that affect helping relationships.

Acknowledgments

I cannot adequately express my appreciation of the reviewers. Their thoughtful comments are reflected in the modifications throughout the text. My sincere thanks to Mark McCabe, Pima Community College; Tricia McClam, University of Tennessee; Roger McNally, State University of New York at Brockport; Frank L. O'Dell, Cleveland State University; and Lew Zachary, Minneapolis Community College.

I also want to thank my students, colleagues, and personal friends for their invaluable comments and recommendations. I am indebted to my editor, Claire Verduin, for her continued support and encouragement, and it has been a delight to work again with production editor Penelope Sky and to benefit from the work of Lorraine Anderson, a superb copy editor.

As always, my husband Sherman is my most supportive and critical editor. His enthusiasm and generativity provide the impetus for my professional and personal development. And I'm fortunate to receive ongoing teaching about human relations from my children and their spouses: Marcia, Joshua, Jeffrey, Alison, and Douglas. They've grown up with this book, and it is indeed gratifying to experience both the process of growth and the outcome, for all of us.

—Barbara F. Okun

 # Contents

ONE

Introduction   **1**

The Purpose of This Book   3
Who Is the Helper?   5
What Makes a Helper Successful?   7
The Two Stages of Counseling   10
The Human Relations Counseling Model   11
Summary   17
References and Further Reading   18

TWO

The Helping Relationship   **20**

Kinds of Helping Relationships   21
How Helping Relationships Develop   22
Effective Communication Behaviors   23
Characteristics of Effective Helpers   30
Helper Self-Assessment   40
Client Variables   42
Summary   42
Comments on Exercises   43
References and Further Reading   45

THREE

Communication Skills   47

Perceiving Nonverbal Messages   47
Hearing Verbal Messages   51
Responding Verbally and Nonverbally   59
Summary   73
Exercise Answers   74
References and Further Reading   77

FOUR

Stage 1: Building the Relationship   78

Conditions Affecting the Relationship Stage   78
Step 1: Initiation/Entry   82
Step 2: Clarification of Presenting Problem   85
Step 3: Definition of Structure/Contract   87
Step 4: Intensive Exploration of Problems   90
Step 5: Establishment of Possible Goals
    and Objectives   91
Resistance   93
Summary   100
Exercise Answers   101
References and Further Reading   101

FIVE

Helping Theory   102

Personal Theories of Human Behavior   102
Psychodynamic Theory   104
Phenomenological Theory   111
Behavioral Theory   117
Cognitive-Behavioral Theory   119

Transactional Analysis Theory   124
Integrative Theories   126
The Systems Perspective   129
Summary   130
References and Further Reading   134

SIX

Introduction to Strategies   138

Strategies and the Three Main Problem Areas   138
Affective Strategies   142
Affective–Cognitive Strategies   148
Cognitive Strategies   150
Cognitive–Behavioral Strategies   152
Behavioral Strategies   160
Strategies That Cut across Domains   169
Summary   179
Exercise Answers   180
References and Further Reading   181

SEVEN

Stage 2: Applying Strategies   185

Step 1: Mutual Acceptance of Defined Goals
    and Objectives   185
Step 2: Planning of Strategies   188
Step 3: Use of Strategies   189
Step 4: Evaluation of Strategies   190
Step 5: Termination   192
Step 6: Follow-Up   196
Case Studies   199
Summary   211
Exercise Answers   211

References and Further Reading   211

EIGHT

Crisis Theory and Intervention   212

What Is a Crisis?   212
Kinds of Crises   214
Who Deals with Crises?   215
Crisis Theory   216
Crisis Intervention   218
Summary   230
References and Further Reading   230

NINE

Issues Affecting Helping   232

Personal Values   232
Ethical Considerations   249
Other Issues That Affect the Counseling Process   253
Common Problems You May
    Encounter as a Helper   257
Recent Trends   258
Summary and Conclusions   260
References and Further Reading   261

Glossary   264

**Appendix A: Observer's Guide to Rating
Communication Skills**   274

Appendix B: Sample Psychotherapy Policies   277

Index   279

**FOURTH
EDITION**

Effective Helping
Interviewing and Counseling Techniques

ONE

Introduction

Since the first edition of *Effective Helping* was published in the mid-1970s, the climate in which people are helped has changed significantly.

The 1970s and 1980s were decades during which females (and then males) became more aware of and articulate about their core need to relate to others. Contemporary psychological theory, influenced by the feminist movement, began to articulate how interpersonal connectedness helps alleviate the alienation, fear, and loneliness associated with rapid societal changes.

During that time period, helping professionals added much to our knowledge of therapy and counseling. We now have additional tools, such as hypnosis, biofeedback, and newer drugs, to deal with some major illnesses and anxieties. We also have better assessment tools, computerized programs, and strategies to help people to manage situational and developmental stress and to enhance their competencies, coping skills, and resources so that they can more effectively negotiate relationships and life transitions. We are beginning to incorporate some of the positive social developments—represented by greater awareness of sexual, racial, and ethical issues—into the delivery of helping services in the United States. And, just as important, there seem to be significantly greater public awareness, acceptance, and expectations of these services.

Now, early in the 1990s, conditions generating psychosocial stress both complicate and add to the need for helping services. The economic downturn at the end of 1990 resulted in increasing unemployment, with significant financial, personal, and family strains. Call-ups of military reserves, followed by combat deployments, created crisis conditions for many of those left behind, who, not surprisingly, were ill-prepared to deal with this type of trauma. Underlying these gloom-generating developments are concerns about matters such as the vast number of homeless (many of whom have emotional/mental problems and in simpler times would have been institutionalized); the growing AIDS epidemic and increased mortality rates from some diseases, despite advances in knowledge and treatment; extensive drug abuse and violence, indicative of the breakdown of social values and structures; questions about the viability of our health care system;

1

and erosion of confidence in our business, educational, health, and political institutions.

And the changes continue, bewildering and confusing us. The global communication revolution—represented by satellite transmissions, computers, fax, videos, car phones—has given us instant access to each other, if not real communication and understanding. We become quickly aware as governments fall, new alliances form, and frontiers alter. While new freedoms and expanded choices are available to us all, they pose bewildering dilemmas of individualism versus group or community cohesion. In this country, the erosion of basic institutions—schools, government, church, and family—in an environment of massive, overwhelming amounts of information places further strains on us all. Perhaps these conditions are one factor contributing to political extremism and the intensity of "one issue" advocates.

Furthermore, today we live in an increasingly multicultural society as a result of immigration and demographic changes. No wonder there is no unified heritage to provide societal consensus about priorities and norms, and loyalty to family and communities. And, as previous norms are overturned and the security found in traditional expectations and lifestyles is lost, we are faced with an ever-higher incidence of individual and family distress related to poverty, poor housing, infectious diseases, homelessness, crime, substance abuse. Our occasional nostalgia for the "good old days"—and mixed envy of the slowly changing traditional societies—is understandable.

The need and desire to care about and help one another still continue to develop in the 1990s, but are increasingly restricted by economic and political factors that have increased competition for drastically limited resources and resulted in cutback and elimination of helping professionals and human services. For example, "managed health care," which is rapidly replacing traditional health insurance plans, provides psychotherapeutic time-limited crisis intervention services that are more dependent upon medications and impersonal short-term service delivery than on helping relationships tailored to individual need. Community agencies have reduced funds and need insurance reimbursement for a significant proportion of the population they serve. The disparity between the increasing need of our stressed population for human services and the decreased availability of and support for such services from an insurance-payment perspective is indeed paradoxical.

The increase in societal complexity makes it essential to consider individuals and their behaviors within the psychosocial contexts of their relevant social and cultural systems: varying styles of immediate and extended families, ever-changing neighborhoods and communities, and both conventional and alternative work and school settings. Not only is it important to differentiate among the particular personal, interpersonal, cultural, and societal variables contributing to an individual's stress level, but it is also necessary to understand the interrelationship of those variables. For example, fear of losing your job in a constricted economy or dissatisfaction with your job may make you feel trapped and frustrated. That frustration may surface in marital arguments or in physical symptoms such as headaches, ulcers, or hypertension. Usually there are multiple external and internal contributing factors to personal problems.

The mass media have drawn attention to the many different forms of today's social ills, and we know that personal, familial, and environmental stresses can show up in physical, psychological, and social symptoms. Underlying or at the very least contributing to these symptoms are interpersonal difficulties, which affect friendships, familial and work relationships, and community, national, and international relations. Most of us feel sincere concern for others and their problems despite our own personal struggles for survival, but we still experience misunderstandings resulting from our inability both to communicate that concern and desire to help and to focus on specific problems and issues of concern. Particularly in this era of budget-constrained service delivery, we need to learn how to communicate our concerns clearly, effectively, and expediently to each other. And as helpers we must develop others' capacities to communicate effectively.

Ineffective or faulty communication is at the root of most interpersonal difficulties. Conversely, effective communication is necessary to develop and to maintain any type of positive interpersonal relationship. Unfortunately, only written communication skills, to the exclusion of face-to-face or interpersonal communication skills, have been traditionally regarded as legitimate curriculum concerns in our schools. Although schools teach us to respond to information, they do not really encourage or teach us to hear, perceive, and respond to more subtle, underlying verbal and nonverbal messages.

Communication skills awareness and training are essential for any human relations endeavor, regardless of the agency or institutional context, regardless of whether the helping relationship is to be short-term or ongoing. *Effective Helping* is intended, in part, to provide a vehicle with which you can develop these skills in order to increase your own and others' self-awareness, understanding of social forces, and capacities for problem solving.

☐ The Purpose of This Book

The basic purpose of this book is to provide a foundation for individuals to develop the human relations skills they need to build effective helping relationships. As part of this foundation, an introductory overview of the counseling process is presented to familiarize helpers with the knowledge and skills used for immediate, short-term, and long-term helping.

The major premise of this book is that *every* individual can learn more effective communication skills that can be applied in personal, social, occupational, and professional settings. Effective communication is the core of the helping process and allows for more satisfying relationships of all types. Improved interpersonal relations allow people to seek and receive support as well as give it to those experiencing personal, familial, occupational, or social distress.

The book is intended for use by both groups and individuals, in both formal and informal human relations training. It focuses on the knowledge and skills needed by individuals in human services positions (such as mental health assistants, counselors, probation officers, employment service workers) or persons involved in other helping roles (such as marriage partners, friends, supervisors, teachers, colleagues).

The material is designed for use in training either beginning students entering helping professions or those people who need or want to improve their human relations effectiveness. Because it teaches fundamental skills, this material will be useful to people continuing on for professional training in counseling as well as to those in nonprofessional settings who find themselves in informal helping relationships in their day-to-day encounters. Although some of the examples and case studies emphasize professional helping skills, they are useful to all helpers in that they illustrate the material in the text and show what goes on in professional helping relationships. The book sometimes uses technical terms with which you may be unfamiliar. The first time each of these terms appears it will be in **boldface** type to indicate its inclusion in the glossary at the end of the book.

Overall, the book is intended to be a practical, applied-skills manual rather than a theoretical treatise. However, it does include a basic overview of current major theoretical approaches to helping, as a background for understanding the strategies (applications of theories) covered. It is an introduction to applied human relations skills in which users are encouraged to employ their own knowledge, learn from their own experiences, and integrate new knowledge with their own capabilities. Remember, though, that human relations—the interactions among people—is a vast subject. Only a limited understanding of this field can be gained from a book such as this. You cannot expect to become expert in this subject from the introductory exposure to theory, skills, and practice that can be given in one book.

The approach to helping presented in this book is flexible and adaptable: whatever strategies or techniques are considered most reasonable and useful in a given situation will be applied, rather than one or only a few theoretical modalities for all helping situations. The strategies that work for a particular client may be modified or rejected for another client in the same or another situation. Likewise, certain strategies will be more compatible with the personal values and style of the helper than will others.

Even though you will not be able to use the strategies presented in this book without further training or assistance, you can begin to use the communication skills covered. Also, your knowledge of the strategies and their applications will help you understand the part that counseling plays in the delivery of human services and relate your work to that performed by professional counselors in the human services field. For example, if you are working as a probation officer and find that one or more of your charges has difficulty meeting the terms of probation, it could be helpful and important for you to know about behavior modification strategies and how they can help your clients gain some control over their environments and behaviors. You might need some assistance in formulating and applying those strategies, but you would at least know which areas of knowledge and training you need to develop.

The overall view of this book is that (1) effective communication is the core of every helping relationship; (2) the goals of every helper include assisting the helpee to (a) increase self-esteem and achieve self-acceptance and (b) gain control over and assume responsibility for his or her behavior and decisions; (3) more than one strategy can be used with any client; (4) continual self-evaluation by the helper and evaluation of "where the helping relationship is" are necessary for effective helping; (5) the helper must be aware of his or her own values, feelings, and thoughts to be able to accept helpees for their own needs, not for those of the helper; and (6) the helper must be sensitive to the gender and cultural bases of clients' beliefs, values, and behaviors.

☐ Who Is the Helper?

The helper is anyone who assists others to understand, overcome, or deal with external or internal problems. We often think of human relations helpers as trained specialists: psychiatrists, psychologists, social workers, psychiatric nurses, or counselors. But an increasing variety of human services workers provide direct or indirect case management and counseling services to a broad array of clients in different private and public settings. Some of these are professional and some are paraprofessional, working adjunctive to or independent of professional helpers. Then there are informal helpers—such as friends, volunteers, relatives, neighborhood workers—who formally or informally find themselves in helper roles. These categories of helpers are not mutually exclusive, and in fact can and do overlap (see Table 1.1).

What distinguishes these three categories of helpers from one another is the level of skills and knowledge they possess.

Table 1.1 Skills and knowledge required by helpers in three categories

	Nonprofessional helper	Paraprofessional helper	Professional helper
Communication skills	X	X	X
Developmental knowledge	X	X	X
Assessment skills		X	X

The counseling skills and knowledge covered in this book can apply in different ways to professional, paraprofessional, or nonprofessional helpers. Because the basic communication skills involved in formal or informal helping and in professional, paraprofessional, or nonprofessional helping relationships are the same, much of what constitutes professional training has proven to be effective for paraprofessional and lay helpers.

Professional Helpers

Professional helpers are specialists who undergo extensive graduate-level training in the study of human behavior, learn applied helping strategies, and experience supervised clinical training while helping individuals, families, and groups.

Although there may be a great deal of overlap in the services delivered by trained specialists, they have different training backgrounds and credentialing requirements. Psychiatrists are physicians who have completed residencies in mental hospitals or on psychiatric units of general hospitals. Their unique contribution to the helping professions includes a knowledge of psychopharmacology and an ability to prescribe drugs, familiarity with medical diseases and their treatment, and experience with severely ill populations. Psychologists, on the other hand, receive training (usually at the doctoral level) in behavioral sciences and are particularly well versed in psychological (learning, developmental, and personality) theory as opposed to the medical model. Their unique contribution is in the field of psychodiagnosis and in research methodology. In a third professional category, counselors usually complete a minimum of two years of graduate study with an emphasis on providing preventive, developmental services as opposed to the correction of severe disturbances. They take many of the same graduate courses as do psychologists, but with a concentration in practitioner rather than methodology courses. Social workers also undertake two years of graduate study and, though their courses are more closely aligned with the remediational medical model, they provide unique services through their knowledge and coordination of available community and governmental services.

All of these professionals offer counseling or therapy to clients as individuals, families, and groups, and it is often impossible to distinguish among types of therapy on the basis of professional identity.

Professional continuing education provides a forum for interdisciplinary interaction and exposure to knowledge common to all helping professions. Thus, the similarities and differences among professional helpers may lie more in individual styles and practices than in professional identities. In the 1990s, however, economic constraints are causing increased competition among these professional disciplines compared to the integration trend of the 1980s. Unfortunately, this can cause unwarranted claims of specialization and work against effective interdisciplinary teamwork as helpers guard their own professional turf.

Paraprofessional Helpers

Overlapping the professional category of helpers is the category of paraprofessional human services workers such as psychiatric aides or technicians, youth street workers, day-care staff, probation officers, and church workers. They normally receive specialized human relations training at the undergraduate college level and usually work on a team with professionals or have professionals available for consultation and supervision. Much of their training occurs on the job, both formally and informally.

Nonprofessional Helpers

We certainly must include the nonprofessionals in our discussion. Although they probably do not receive formal training as helpers, they may attend seminars or meetings on various issues in human relations services from time to time. This group includes people who provide important helping assistance on a formal basis (interviewers, supervisors, teachers), on a semiformal basis (volunteers), and on an informal basis (friends, relatives, colleagues).

The common denominator of the three groups of helpers is that they all must effectively use communication skills to initiate and develop helping relationships with the people whom they are assisting. To provide support for extensive kinds of problems, helpers apply certain strategies. The application of those strategies requires formal training and experience; this book illustrates their use by professional and some paraprofessional helpers.

☐ What Makes a Helper Successful?

The successful helper is familiar with many approaches and strategies. Having a broad range of alternatives enables helpers to select those strategies most likely to meet the needs of a particular client or client system. When the selected strategies are applied, they are filtered through the unique personality of the helper. In other words, each person's perceptions, attitudes, thoughts, and feelings affect his or her interpretation and application of the theory. In fact, it is often said that the personal attributes of the helper are more important than strategic skills.

Underlying the effectiveness of any given strategy is the level of trust between helper and helpee that is developed during the first stage of helping, the relationship stage. Trust is developed using communication skills within an empathic context. **Empathy**, defined as understanding another person from that person's frame of reference, is vital to the effectiveness of communication skills. Empathic communication skills leading to trust, then, are essential to the effectiveness of the whole helping process.

Helpers need to remember that there are cultural differences in the ways groups express empathy; in other words, what may be empathic behavior for one helpee (for example, touching) may *not* be so for another. Thus, selection of strategies will be influenced by cultural factors. While there are some basic counseling skills and strategies that cut across class, race, and culture, helpers must adapt their counseling style to achieve congruence with the value systems of culturally diverse clients. Sensitivity to the nuances and implications of cultural variables is necessary if one is to be effective with clients from a variety of backgrounds.

To be comfortable applying a variety of helping strategies, the helper must be able to deal with others in the **affective** domain (relating to feelings or emotion), the **cognitive** domain (relating to thinking or intellectual processes), and the **behavioral** domain (relating to action or deeds). By extension, the helper must teach the client to function more effectively in all three domains. Therefore, helpers must continually develop an understanding of themselves; they need to clarify their own social, economic, and cultural values in order to recognize and separate their needs and problems from those of their clients. The strategy selected for formally helping a particular client may depend on the helper's assessment of deficits in a particular domain (affective, cognitive, or behavioral) as well as the helper's theoretical perspective.

The following examples demonstrate these points.

A client was once referred to me by a colleague specifically for **systematic desensitization**, a precise behavioral technique developed by Joseph Wolpe and aimed at reducing anxiety by associating the undesirable response with relaxation, an antagonistic response to anxiety, to extinguish the undesirable response. This client was a 30-year-old male who experienced rage at what he considered undue noise: His wife's chewing, his co-workers' pencil tapping, his baby's crying all enraged him. After several sessions, we determined that the client was functioning for the most part in the cognitive and behavioral domains and was completely detached from any of his own or anyone else's feelings. Thus he was unable to experience any effective human relationships and his marriage was in danger. During the following few sessions we unsuccessfully attempted systematic desensitization to satisfy the client's expectations. At the same time, I endeavored to establish a trustful relationship. When he began to trust me and to feel more comfortable, I suggested that we use some **client-centered** and **Gestalt** strategies to tap his affective domain, to help him become aware of and explore his feelings. The client's wife was eventually brought into the counseling sessions, and she verified his reports that, as he began to learn to experience and explore his emotions, he was better able to relate to her and others and his tolerance of outside noises began to increase.

A young lady was babysitting for a family of three children while the parents were away on a trip. A recently widowed grandmother had moved in with this family and was also in the house. The babysitter had been told by the parents that the grandmother was in the depressed stage of a manic-depressive illness. During the first two days, the babysitter observed that Grandma controlled the family by refusing to eat, by talking about her poor self, and by continuous moping. Grandma was miserable, and the whole family always felt guilty and could never cheer her up, no matter what they did! Familiar with behavioral principles and reality-therapy techniques and unable to endure the oppressive climate in this house, the babysitter pointed out to the three children how they were feeding this situation, and she modeled the kind of reinforcing behavior she wished them to adopt: ignoring Grandma's complaints and refusal to eat, but sitting down and talking to her and giving her lots of **strokes** (loving attention) when she expressed interest in anything other than herself. At the same time, the babysitter told Grandma that she was on to her and how she pulled the strings of members of the family. She told her that she would not accept her depression or refusal to eat. This was done in a loving but firm manner. By the end of the week, Grandma was discussing the presidential debates with whoever would listen, taking walks with neighbors, and reading newspapers and books as well as eating adequate meals. She had desperately wanted attention, to feel worthwhile. Now she was learning to get that attention by positive behavior.

A government worker was sent to the outplacement counselor after receiving notice of a layoff. He talked at length about his fears of not finding another job, not being able to support his family and make his house payments. He described his symptoms of anxiety as inability to concentrate, eat, or sleep, and irritation with his family and friends. This was the "worst thing that ever had happened" to him and he could not understand why so many people were losing their jobs and how that could have happened to him. Over and over, he described the details of his job performance, determined to find the "error" that he must have committed leading to this "failure." The outplacement counselor felt that he was focusing on his feelings of failure and self-blame excessively, bordering on hysteria. Thus, the counselor decided to try some **cognitive restructuring**, based on rational-emotive therapy. After several sessions and some homework reading, the government worker was able to correct some of his faulty thinking. He no longer blamed himself for the economic problems of the country, no longer felt that he had personally failed and was, therefore, inadequate. As he changed his thinking, his symptoms decreased and he was able to understand his fear and dismay, while taking active steps to prepare and send out resumes, attend sessions about job seeking, and go on actual interviews.

We can see from these examples that (1) different people need help in different areas of functioning; (2) outcomes are more likely to be successful if helpers fit the strategy to the helpee's individual needs than if they apply the strategy in the same way all the time; and (3) sometimes the effective strategy is relatively simple and can be used by someone lacking lengthy professional training.

☐ The Two Stages of Counseling

The term *counseling,* as used in this book, encompasses the professional, para-professional, and nonprofessional forms of helping. The terms *counselor* and *helper* will be used interchangeably, as will the terms *helpee* and *client.*

Many people consider counseling to be both an art and a science. It is an art in the sense that the personality, values, and demeanor (along with the skills and knowledge) of the counselor are subjective variables in the counseling process that are difficult to define or measure. It is a science in that much of what we know about human behavior and some of the helping strategies has been developed into structured, measurable, objective counseling systems. Counseling can be thought of as a process with two overlapping parts or stages: the first stage more an art, the second more a science. And the counselor's style of delivery is perhaps an art that is practiced throughout the entire helping relationship.

The first stage of the helping process focuses on building rapport and trust between the helper and the helpee. The helper offers the helpee support for self-disclosure to uncover and explore as much information and as many feelings as possible and pertinent. This exploration enables the helper and the helpee to focus on the helpee's needs and presenting issues and to determine mutually the goals and objectives for helping and, thus, the direction of the helping relationship.

The skills involved in relationship building on a one-to-one (that is, one helper, one helpee) basis are fundamental skills that can be used when interacting with others at home, at school, at work, or in the community. These relationship skills have been identified in the work of Carkhuff (1967, 1969, 1971, 1983, 1986), Egan (1990), Gordon (1970), Kagan (1980, 1984), Ivey (1988), Ivey and Authier (1978), Ivey and Ivey (1990), Ivey, Ivey, and Simek-Downing (1987), and others who have developed helper-training systems that derive from basic Rogerian client-centered theory (which we will discuss in Chapter 5). These systems include listening, attending, perceiving, and responding as components of communication and allow for exploration, clarification, and assessment of the helpee's problems.

As a relationship and problem assessment are developed, the second stage of the helping process begins. This stage comprises strategy planning, implementation, and evaluation, which lead to termination and follow-up. Normally this stage of the helping process is the province of professional helpers, although it is also of some concern to paraprofessional helpers. Although nonprofessional helpers are not usually involved in this stage of helping, they still need a rudimentary knowledge of the theory and application of helping strategies in professional and paraprofessional helping relationships to understand and appropriately use human services resources. The success of the second stage depends greatly on how effective the communication skills were in establishing a positive helping relationship during the first stage.

☐ The Human Relations Counseling Model

This book is based on the human relations counseling model. This model derives from the major formal theoretical views discussed in Chapter 5. It emphasizes a client-centered, problem-solving helping relationship in which behavior changes and action can result from one or both of the following: (1) the client's exploration and understanding of his or her feelings, thoughts, and actions, or (2) the client's understanding of and decision to modify pertinent environmental and systemic variables. Cognitive, affective, or behavioral strategies are used alone or in concert when both the helper and the helpee determine the appropriate need and timing. And some strategies combine various aspects of several formal theories of helping.

Assumptions and Implications of the Model

The theoretical assumptions of the human relations counseling model reflect **existential** as well as behavioristic influences. These assumptions are as follows:

1. People are responsible for and capable of making their own decisions.
2. People are controlled to a certain extent by their environment, but they are able to direct their lives more than they realize. They always have some freedom to choose, even if their options are restricted by environmental variables or inherent biological or personality predispositions.
3. Behaviors are purposive and goal-directed. People are continuously striving toward meeting their own needs, from basic physiological needs to abstract self-actualization (psychological, sociological, and aesthetic) needs.
4. People want to feel good about themselves and continually need positive confirmation of their own self-worth from **significant others.** They want to feel and behave **congruently**, to reduce **dissonance** between internal and external realities.
5. People are capable of learning new behaviors and unlearning existing behaviors, and they are subject to environmental and internal consequences of their behaviors, which in turn serve as **reinforcements.** They strive for reinforcements that are meaningful and congruent with their personal values and belief systems.
6. People's personal problems may arise from unfinished business (unresolved conflicts) stemming from the past (concerning events and relationships) and, although some exploration of causation may be beneficial in some cases, most problems can be worked through by focusing on the here and now—on what choices the person has now. Problems are also caused by **incongruencies** between external and internal perceptions in the present—that is, discrepancies between a person's actual experience and his or her picture of that experience.

7. Many problems experienced by people today are societal or systemic rather than interpersonal or intrapersonal. People are capable of learning to effect choices and changes from within the system as well as from without.

Exercise 1.1   Go back over the seven assumptions just listed. Do you find that you strongly disagree, disagree, agree, or strongly agree with each assumption? How would you change each assumption and why? What would you add? Which assumption do you have the most difficulty accepting, and how would that affect how you work with people? As people in your group share their agree-disagree statements, identify those with whom you agree or disagree on each one. You may want to have small-group discussions of each assumption, dividing into strongly disagree, disagree, agree, or strongly agree groupings; reconvene as a whole to discuss the similarities and differences you discover.

The human relations counseling model also emphasizes the mutual identification by the helper and the helpee of goals, objectives, and intervention strategies whose success can ultimately be evaluated according to the observable behavioral change in the helpee. The model represents an eclectic approach in that it uses a variety of counseling techniques and strategies to effect change, but the major vehicle for change remains the development and maintenance of a warm, personally involved, empathic relationship.

The helper is encouraged to learn about the systems (contexts) in which helpees live and function. In addition, the helper is encouraged to learn when and how to use different techniques and strategies and to use different approaches with the same helpee in order to deal with as many areas of concern as possible within the cognitive, affective, and behavioral domains. The goals of helping are to integrate those three domains, to aid the helpee to become emotionally and cognitively aware of his or her responsibilities and choices, and to see that awareness translated into action. When helpees are able to assume responsibility for their feelings, thoughts, and actions and to reduce the contradictions among them, they are able to feel good about themselves and about the world and make choices that reflect the integration of internal and external variables. They are then able to behave proactively rather than reactively in their relationship systems.

As previously stated, the helping relationship is considered to be the essential foundation of the helping process. And it is the **process** of verbal and nonverbal communication, not the content, on which that relationship is based. As long as there is an effective helping relationship that communicates to the helpee the helper's capacity for understanding, humanness, and strength to resist manipulation, there is a safe, protective environment allowing flexibility in selection and use of strategies.

Strategies are secondary to the helping relationship. In fact, research indicates that client variables and counselor variables are more significant than technique variables in the helping process. (Helper and helpee variables will be elaborated in Chapter 2.) If a particular strategy does not work but the helping relationship is solid, the helping process is not likely to be negatively affected. For example, if you have developed a trustful relationship with a helpee and you ask her to *dialogue* (Gestalt technique: see Chapter 5) with her mother—taking both roles to become more aware of her positive and negative feelings toward her mother— and she is unable to do it, she will not think you are crazy or incompetent for having tried this strategy. If she trusts and respects you, she will continue to explore with you, seeking strategies that will be more helpful. This type of helping relationship is reciprocal, in that the helper is considered an equal of the helpee rather than an expert or a magician. "Equal" in this sense means that social distance is minimal and the responsibility for what occurs is mutual; both people work together toward achieving agreed-on objectives. At the same time, the helper must be able to communicate to clients an understanding of human behavior and have the skills to help clients change their behaviors. The helping relationship serves to increase the helpee's self-understanding and self-exploration, but it does not provide false reassurance and support. Rather, it is honest and allows for expression of the discomfort and pain that may be involved in the helping process. This honesty enables helpers to tolerate their own and the helpees' discomfort without needing to cover it up with false reassurance and distancing.

The major implications of the human relations counseling model for helpers are that it

1. Defines empathic communication skills as the core of effective human relationships
2. Allows that empathic communication skills can be taught to all helpers in all types of helping relationships
3. Provides room for diversity and flexibility so helpers can learn a variety of intervention strategies that can be effective if a successful helping relationship is developed and maintained
4. Modifies and integrates a variety of established approaches and strategies
5. Provides the versatility and flexibility necessary to meet the needs of a multicultural heterogeneous population
6. Provides for dealing with feelings, thoughts, and behaviors in a short-term, practical manner relating to the helpee's life
7. Focuses on the positive rather than negative aspects of the helpee's life (that is, on those aspects one can change rather than those over which one has no control)
8. Assists the helpee to actively assume responsibility for living and for decision making

Dimensions of the Model

The human relations counseling model has three integrated dimensions: stages, skills, and issues (see Figure 1.1). Outlining a counseling model in diagrammatic form necessitates some formalizing and systematizing that seem rigid and arbi-

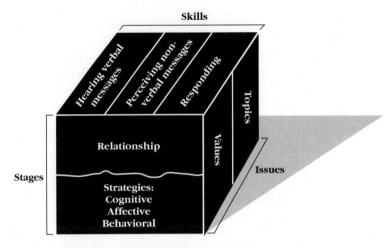

Figure 1.1 The counseling model in dimensional terms

trary. However, this multidimensional view is useful in conceptualizing what happens in and what constitutes effective helping. It thereby provides a useful framework for learning about counseling and developing necessary skills. Naturally, the helper will modify or redesign this conceptual model into whatever form works for him or her.

The First Dimension

The first dimension comprises the two stages of the helping process described earlier. These two stages consist of the following steps:

1. Relationship (development of rapport, trust, honesty, empathy)
    a. Initiation/entry
    b. Identification and clarification of problem(s) being presented
    c. Agreement on structure/contract for helping relationship
    d. Intensive exploration of problem(s)
    e. Definition of possible goals and objectives of helping relationship
2. Strategies (work)
    a. Mutual acceptance of defined goals and objectives of helping relationship
    b. Planning of strategies
    c. Use of strategies
    d. Evaluation of strategies
    e. Termination
    f. Follow-up

The thesis of this book is that the development of a warm, trustful relationship between the helper and the helpee underlies any strategy or approach to the helping process and therefore is a basic condition for the success of any helping process. Naturally, this relationship depends on one's theoretical view of people, behavior, the world, and helping. Developing a relationship is a time-consuming process; however, a skilled helper can guide this development so that the relationship can aid the helpee within a short period of time. This is critical for short-term, problem-focused, managed service delivery.

Development starts with the initial contact between the helper and the helpee. A climate is provided for the helpee to explore problems and to begin to identify underlying as well as apparent concerns. Later, the client begins to understand those concerns and their implications for living and starts to clarify his or her needs and expectations of the helping relationship to facilitate self-exploration, self-understanding, and choices of action. The success of the helping relationship is crucial to the mutual determination of appropriate goals and objectives.

Once the goals and objectives have been mutually decided on, the helper reviews all available effective strategies (or courses of action for effective helping) and discusses with the helpee the rationale for choosing a particular strategy. The possible consequences and ramifications of any strategy are explored.

When agreement on a course of action is reached, the helper applies the strategy, keeping his or her mind open to modifying or refining it, depending on the needs of the helpee. Evaluation of the effectiveness of the chosen strategy must be continual.

When the outcomes agreed on by both helper and helpee have been achieved, either the helping relationship is terminated or attention is focused on another set of objectives and goals. If the relationship is terminated, the helper later informally or formally checks up on the progress of the helpee. Ideally, termination is a process that occurs over time rather than by a sudden cessation of activity. Ward (1984) conceptualizes three major functions of the termination process: (1) assessing helpee readiness for the end of the helping relationship; (2) bringing about appropriate closure of the helping relationship; and (3) maximizing the helpee's self-reliance and confidence to maintain change after the helping relationship has ended. Successful termination implies that the relationship and problem-solving skills the helpee has learned during the helping process will be applied to future relationships and problems. Hence, the process of letting go and achieving closure in relationships is just as important as the process of developing a new relationship. Unfortunately, time-limit pressures sometimes cause the helpers to overlook this important process.

The Second Dimension

The second dimension of the counseling model represents communication skills: hearing verbal messages, perceiving nonverbal messages, and responding to verbal and nonverbal messages. *Verbal messages* are the apparent and underlying cognitive and affective content of the helpee's statements. Understanding the implicit and explicit content is usually secondary in importance to understanding the feelings communicated by the helpee. *Nonverbal messages* are conveyed by body language, vocal tone, facial expression, and other cues that accompany verbal messages. The helper learns to recognize inconsistencies between verbal and nonverbal messages and to develop the helpee's awareness of those inconsistencies. *Responding* requires immediate, genuine, concrete, and empathic reaction to verbal and nonverbal messages. Both the apparent and underlying significances of messages as well as their relationships and inconsistencies determine appropriate responses.

These communication skills are required to effect the two stages of helping that constitute the first dimension. The model assumes consistency between the helper's verbal and nonverbal messages. It also relies on the helper's ability to respond to the helpee by clarifying the latter's underlying feelings and thoughts in such a way as to increase the helpee's self-understanding.

By developing communication skills, helpers also develop their own self-awareness. As they learn to use their intuitive feelings as guidelines for hearing other people's messages, they sharpen their helping skills. Helpers are always asking themselves, "What is this person really trying to say to me?" "What is she or he really feeling?"—and they try to communicate their understanding of the message and feeling back to the helpee.

The Third Dimension

The third dimension of the counseling model represents issues, which are the values and cognitive topics that cut across the other two dimensions. These issues involve not only how an individual relates to others and to his or her environment, but also such subjects as sexism, racism, ageism, and poverty. Furthermore, this dimension includes professional matters of ethics, training, and practice, as well as the personal values and attitudes of the helper.

Pervasive issues affect both stages of helping. By exposing and clarifying these issues, helpers are able to achieve the type of helping relationship in which they do not interfere with helping. Responsive listening skills are effective techniques for discovering and exploring such issues.

To effect values clarification, the helper and the helpee both must take responsibility for their own attitudes, beliefs, and values. For example, a counselor who tells a female client that she should not think about returning to work until her child is in school is allowing his or her sexist values to distort or interfere with counseling. If helpers are not aware of their own biases, the effects will be harmful. However, if they recognize their biases, they will be less likely to impose them on clients. Research has shown that helpers do communicate their values to helpees, whether or not they do so consciously. Bringing those values into the open and being constantly aware of them can keep helpers from imposing them on others.

Exercise 1.2   With a partner or in a small group, discuss what thoughts and feelings you might experience as a helper counseling the following clients: (1) a 14-year-old pregnant girl who wants an abortion without her parents' knowing; (2) a married man who confides that he has had several homosexual extramarital affairs; (3) a 26-year-old man living at home with his parents who refuses to get a job or take responsibility for himself; (4) a recently separated mother who wants to have her drug-addicted boyfriend move in with her and her young children so that she can "help him" get better. How do you think your true thoughts and feelings might influence the counseling process? What kinds of things would you want to say and do? What differences emerge in your group and how do you feel about them? What have you learned about yourself and others?

Some of the current concerns affecting the helping process include how to help involuntary or reluctant clients, how to help people we really dislike, and how to deal with complex ethical issues such as informed consent, confidentiality, and abuse of power. Also, the helper's responsibility to the sponsoring institution can lead to conflicts of interest and ethical infractions like keeping clients longer than necessary to maintain a census count and/or unrealistically trying to squeeze services into a limited time frame. The question becomes: Whose needs take priority—the helper's, the helpee's, or the institution's?

The human relations counseling model will be discussed throughout the book. Chapter 2 defines and illustrates effective and ineffective helping relationships. Chapter 3 presents materials for developing techniques for effective communication. Chapter 4 explores the relationship stage in depth, and Chapter 5 presents an overview of theoretical approaches that relate to the strategies discussed in Chapter 6. In Chapter 7 the application of strategies is explored, and in Chapter 8 crisis theory and crisis intervention are presented. Chapter 9 gives a brief overview of issues affecting the helping process and a final postscript summarizing the entire model.

Interspersed within these chapters are case materials and exercises designed to provide you with an opportunity to use your conceptual and practical understanding of the material covered in the text. The exercises are intended to be used primarily in supervised group settings.

☐ Summary

The intent of this book is to provide a fundamental introduction to the skills and knowledge necessary for effective helping relationships. These skills and knowledge are needed, to varying degrees, by nonprofessional, paraprofessional, and professional human services workers to develop and maintain satisfactory and helpful interpersonal relationships.

We began this chapter by describing the impact of technology and complex social change on individuals and families. The confusion and problems emanating from those changes can heighten anxiety and feelings of alienation and helplessness: hence our focus on how individuals can help themselves and others to feel less alienated and powerless by improving their own interpersonal relationships. If helpers achieve quality interpersonal relations in their own lives, they can model their skills in the helping relationship and teach others to improve the quality of relationships. Counseling, one type of helping interaction and an important part of human services, is used to demonstrate the helping relationship.

The purpose of the helping relationship was defined as assisting clients to achieve greater self-acceptance and self-esteem and to gain control over their behavior and decisions. The relationship is based on the communication of empathy and application of appropriate strategies. Thus, the human relations counseling model consists of three equally important and interdependent dimensions: stages (relationship and strategies), skills, and issues. The helping process depends absolutely on the development of a trustful relationship between helper and helpee. Effective communication skills enhance that relationship and also provide a way for dealing with controversial issues, while strategies are the various approaches that helpers use to promote self-exploration, understanding, and behavior change in helpees, which in turn lead to heightened self-acceptance and responsibility. The strategies aim to increase the helpee's awareness and successful functioning in the affective (feeling), behavioral (doing), and cognitive (thinking) domains. Termination occurs when both helper and helpee feel that the helpee has resolved and worked through his or her concerns and can apply what has been learned from the helping relationship to future situations and relationships.

☐ References and Further Reading

Carkhuff, R. (1967). *The counselor's contribution to facilitative processes.* Urbana, IL: Parkinson.

Carkhuff, R. (1969). *Helping and human relations* (Vols. 1 and 2). New York: Holt, Rinehart & Winston.

Carkhuff, R. (1971). *The development of human resources.* New York: Holt, Rinehart & Winston.

Carkhuff, R. (1983). *The art of helping.* (5th ed.). Amherst, MA: Human Resource Development Press.

Carkhuff, R. (1986). *Human processing and human productivity.* Amherst, MA: Human Resource Development Press.

Carkhuff, R., & Berenson, B. (1967). *Beyond counseling and therapy.* New York: Holt, Rinehart & Winston.

Carkhuff, R., & Berenson, B. (1976). *Teaching as treatment.* Amherst, MA: Human Resource Development Press.

Corey, G. (1990). *Theory and practice of counseling and psychotherapy.* (4th ed.). Pacific Grove, CA: Brooks/Cole.

Cormier, W. H., & Cormier, L. S. (1985). *Interviewing strategies for helpers: A guide to assessment, treatment, and evaluation* (2nd ed.). Pacific Grove, CA: Brooks/Cole.

Egan, G. (1990). *The skilled helper* (4th ed.). Pacific Grove, CA: Brooks/Cole.

Gazda, G., Asbury, F., Balzer, F., Childers, W., & Walters, R. (1977). *Human relations development* (2nd ed.). Boston: Allyn & Bacon.

Gordon, T. (1970). *Parent effectiveness training.* New York: Wyden.

Ivey, A. (1971). *Microcounseling: Innovations in interviewing training.* Springfield, IL: Charles C Thomas.

Ivey, A. (1988). *Intentional interviewing and counseling: Facilitating client development.* Pacific Grove, CA: Brooks/Cole.

Ivey, A., & Authier, J. (1978). *Microcounseling: Innovations in interviewing, counseling, psychotherapy, and psychoeducation.* Springfield, IL: Charles C Thomas.

Ivey, A., & Ivey, M. (1990). Assessing and facilitating children's cognitive development: Developmental counseling and therapy in a case of child abuse. *Journal of Counseling and Development, 68,* 299–305.

Ivey, A., Ivey, M., & Simek-Downing, L. (1987). *Counseling and psychotherapy: Integrating skills, theories, and practice.* Englewood Cliffs, NJ: Prentice-Hall.

Kagan, N. (1980). *Interpersonal process recall: A method of influencing human interaction.* Houston: Mason Media.

Kagan, N. (1984). Interpersonal process recall: Basic methods and recent research. In D. Larson (Ed.), *Teaching psychological skills: Models for giving psychology away.* Pacific Grove, CA: Brooks/Cole.

Rogers, C. (Ed.). (1967). *The therapeutic relationship and its impact.* Madison: University of Wisconsin Press.

Ward, D. E. (1984). Termination of individual counseling: Concepts and strategies. *Journal of Counseling and Development, 63*(1), 21–26.

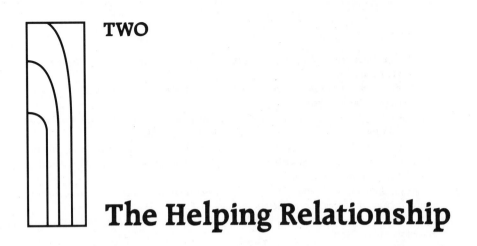

TWO

The Helping Relationship

This chapter provides a description of helping relationships, with particular emphasis on the communication involved. Having an overall view of what occurs in a helping relationship will increase your understanding of subsequent chapters, in which we begin to talk about and practice helping skills and theoretical knowledge and see how they interrelate.

The purpose of a helping relationship is to meet the needs of the helpee, not those of the helper. The relationship is meant largely to enable helpees to assume responsibility for themselves and make their own decisions based on expanded alternatives and approaches. Therefore, helpers neither solve helpees' problems nor reassure them merely to make them feel better.

Helpers assist and support helpees so they can come to terms with their problems by exploration, understanding, and action. For instance, if an employee comes to you and says he or she cannot continue working for a particular supervisor, you may, after exploration, focus on helping the employee learn to get along better in that work setting. On the other hand, helping may well involve some direct form of environmental rearranging or systems change that has the effect of aiding someone in the system. If, in this instance, exploration determines that the supervisor is the source of the trouble and is hampering the employee's contributions to production or service, you may focus on transferring the employee to a more suitable setting, work directly with the supervisor to improve relations, or do both. A helping situation does not involve doing something to someone else to make him or her better; it does involve working together to seek the best solution (after considering all possible alternatives) and, if possible, to implement that solution.

Therefore, a helping relationship that benefits the helpee is a mutual learning process between the helpee and one or more other persons. The effectiveness of the relationship depends on (1) the helpee's skill in communicating his or her understanding of the helper's feelings and behaviors; (2) the helper's ability to determine and clarify the helpee's problem; and (3) the helper's ability to apply appropriate helping strategies in order to facilitate the recipient's self-exploration, self-understanding, problem solving, and decision making, all of which lead to constructive action on the part of the helpee.

☐ Kinds of Helping Relationships

There are different kinds of helping relationships—although all are similar in concept and strategies used—corresponding to the three categories of helpers discussed in Chapter 1. There are professional relationships (doctor/patient, pastor/congregant, counselor/client, social worker/client) for which the helper has received intensive formal training in human behavior, problem solving, and communication related to helping. There are paraprofessional relationships (employment interviewer/applicant, case aide/client, street worker/youth, recreation leader/youth, probation officer/probationer) for which the helper has received short-term formal training such as courses or workshops in human relations. And there are nonprofessional helping relationships (receptionist/customer, salesperson/customer, flight attendant/passenger, volunteer/client) in which the helping process may be incidental to the relationship.

Within these three categories a distinction can also be made between formal and informal relationships (see Figure 2.1). Formal situations are ones in which the helper/helpee roles are stated or implied by position or contract and the specific reason for contact is known to be for some kind of help to be given. Informal helping occurs when the helping relationship is secondary to another relationship, whether formal or informal: principal/teacher, friend/friend. Formal relationships usually occur in an institutional setting such as an office, school, or hospital, and informal relationships can occur any place: between friends, relatives, neighbors, or peers, in the office, school, or hospital. One sees less structure,

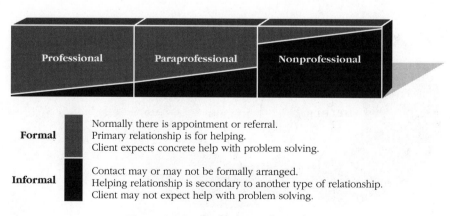

Formal	Normally there is appointment or referral. Primary relationship is for helping. Client expects concrete help with problem solving.
Informal	Contact may or may not be formally arranged. Helping relationship is secondary to another type of relationship. Client may not expect help with problem solving.

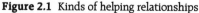

Figure 2.1 Kinds of helping relationships

shorter time involvement, and more limited expectations for problem solving as the relationships become less formal.

☐ How Helping Relationships Develop

Helping relationships begin with a helper and a helpee meeting to focus attention on the helpee's concerns. Thus, a helping relationship is distinct from other relationships in this focus on one party's concerns and issues. However, it shares ingredients common to all satisfactory relationships—ingredients such as trust, empathy, genuineness, concern and caring, respect, tolerance and acceptance, honesty, commitment to the relationship, and dependability. These ingredients are not usually present at the beginning of a relationship but develop over time as people get to know one another. If trust does not develop, the other ingredients will dwindle and the relationship will eventually die. Trust is established when an individual perceives and believes that the other person in the relationship will not mislead or harm him or her in any way.

Figure 2.2 shows the helping relationship in its primary form. The helper and the helpee are always engaged in mutual communication. The principal difference between them is that the helper possesses some skills (expertise) and the helpee possesses some concerns (problems). Put any names in the circles and put the circles in any context; the relationship will remain essentially the same. Note that helper and helpee each come to the relationship with a set of attitudes, needs, values, and beliefs. The degree of congruence between those two sets can affect the relationship either positively or negatively. However, the helper's trustworthiness, empathy, nonjudgmentalness, and tolerance can be communicated effectively so as to lessen the possibility of adverse or nonhelpful outcomes when values, needs, attitudes, or beliefs differ.

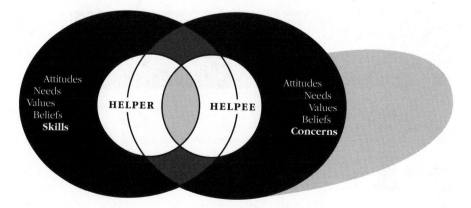

Figure 2.2 The helping relationship in its primary form

☐ Effective Communication Behaviors

Regardless of the setting or nature of the helping relationship, regardless of the personal values and beliefs of the people involved, and regardless of the theoretical orientation of the professional helper, the underlying prerequisite skill in any helping relationship is effective, empathic communication. My experience and a wide spectrum of research verify this conclusion.

How can we help people if we cannot learn their concerns and impart to them our feelings or thoughts? Both processes depend on the ability to communicate. Communication in this sense means the capacity to listen, pay attention, perceive, and respond verbally and nonverbally to the helpee in such a way as to demonstrate to her or him that one has attended, listened, and accurately perceived. It means responding as opposed to reacting. This ability can be learned by most people, whatever their educational background or personality. It is a skill that requires continual practice, as does any other type of skill. Not surprisingly, the people considered by others to be most helpful in formal or informal settings possess good communication skills.

Research indicates that communication problems are the major source of interpersonal difficulties. For example, most marital and family problems stem from misunderstanding, from ineffective communication, which results in frustration and anger when implicit expectations and desires are not fulfilled. And a major problem of those who seek professional help is their inability to recognize and communicate their problems or concerns.

Many people think they know what their problem is, but have difficulty in verbally communicating their concerns. Others are able to verbalize their concerns but need help in identifying the underlying problem. Still others do not even recognize that they have a problem and are what we call "reluctant clients," in that they are required to seek help. In all cases, good verbal and nonverbal communications are essential to both stages of the helping process. Therefore, it is necessary to look closely at the process (a sequence of events that takes place over a period of time) of communication within helping relationships—at the behaviors that encourage and impede communication.

Most of us have been helpees on numerous occasions; so, based on our own experience, we should be able to recognize helpers' behaviors that have aided or hindered our receiving help. If you were asked to list those verbal and nonverbal behaviors that you, as the helpee, found supportive in any kind of helping relationship, what would your list look like? Table 2.1 lists behaviors typically cited by beginning counseling students. Most students think of more verbal behaviors than nonverbal. See if you agree with these lists or can add to them.

The gist of the lists in Table 2.1 is that in the helping relationships students judge most effective, helpers show listening and attending behaviors that communicate empathy, encouragement, support, honesty, caring, concern, respect, sharing, affection, protection, potency, and nonjudgmental acceptance. Clients are helped because they feel worthwhile as human beings, feel accepted by another

Table 2.1   Helpful Behaviors

Verbal	Nonverbal
Uses understandable words	Tone of voice similar to helpee's
Reflects back and clarifies helpee's statements	Maintains good eye contact
Appropriately interprets	Occasional head nodding
Summarizes for helpee	Facial animation
Responds to primary message	Occasional smiling
Uses verbal reinforcers (for example, "Mm-mm," "I see," "Yes")	Occasional hand gesturing
	Close physical proximity to helpee
Calls helpee by first name or "you"	Moderate rate of speech
Appropriately gives information	Body leans toward helpee
Answers questions about self	Occasional touching
Uses humor occasionally to reduce tension	
Is nonjudgmental	
Adds greater understanding to helpee's statement	
Phrases interpretations tentatively so as to elicit genuine feedback from helpee	

human being, and are therefore permitted to be their true selves and to explore their true concerns.

Similarly, you can undoubtedly recognize unhelpful communication behaviors. Would you add to or change the lists presented in Table 2.2? We can see that these verbal and nonverbal behaviors involve inattentiveness, imposition of the helper's values and beliefs on the helpee, judgment, and "I-know-what's-best-for-you" or "I'm-better-than-you" attitudes. These behaviors are hindrances because they put helpees on the defensive immediately and make them feel so worthless that they will naturally choose avoidance rather than approach behaviors.

Exercise 2.1   Generate a list of helpful and unhelpful responses that you personally have experienced in specific situations: for example, when a significant relationship ended, when you received a low grade on a paper or exam, when you didn't get a desired job or into a specific school, when someone close to you moved away, became ill, or died, when you were in an accident, or when you wrecked your car. After you have identified particular responses, ask yourself why they were helpful or unhelpful. In small or large groups, share your findings and discuss why a response may be helpful to one person but not another.

Exercise 2.2   Divide into triads of helper, helpee, and observer. The requisite materials for this exercise are crayons, large drawing sheets, and a blindfold for each triad. Each member of the triad should experience each role. The helpee is blindfolded and helped to draw a picture by the helper. It is important that each helper interpret how best to help in his or her own way. The observer writes down all verbal and nonverbal behaviors and, after each member of the triad has been a helpee, the group works together to produce a list of helpful and unhelpful verbal and nonverbal

Table 2.2   Unhelpful Behaviors

Verbal	Nonverbal
Advice giving	Looking away from helpee
Preaching	Sitting far away or turned away
Placating	from helpee
Blaming	Sneering
Cajoling	Frowning
Exhorting	Scowling
Extensive probing and questioning, especially "why"	Tight mouth
questions	Shaking pointed finger
Directing, demanding	Distracting gestures
Patronizing attitude	Yawning
Overinterpretation	Closing eyes
Using words helpee doesn't understand	Unpleasant tone of voice
Straying from topic	Rate of speech too slow or too fast
Intellectualizing	
Overanalyzing	
Talking about self too much	

behaviors. Again, discuss why a particular behavior was helpful to one person and not to another. What was comfortable and uncomfortable for you as a helper? How did you feel as a helpee? What did you learn about your helping behaviors? What do you think you might want to change?

Exercise 2.3   To identify helpful and unhelpful nonverbal behaviors, take turns playing charades. One helper will verbally communicate an issue or concern or describe a situation. Another helper will nonverbally attempt to help. Observers will identify those nonverbal behaviors that were helpful and those that were unhelpful and discuss their observations with the helpee and helper. The helpee will also give feedback on what was experienced as helpful or unhelpful.

The purpose of Exercises 2.4 through 2.8 is to determine whether or not you can recognize what is effective and what is not effective communication in a helping relationship. Read each of the exercises, ranking the helper's responses on a 0–5 scale (0 = least helpful, 5 = most helpful) before reading the text comments and discussing your fellow students' reactions. Obviously, you can respond only to the verbal content of the interactions. Comments on the exercises are found at the end of the chapter. Please remember that there are usually no right or wrong answers to the exercises in this book. They should be used as a focus for discussion of people's differences in perceptions and responses. If you want additional practice in learning communication skills, you can act out these case interactions in class.

Exercise 2.4   Ms. James is a 26-year-old clerk-typist in the marketing department of a large publishing firm. She has worked there for 16 months, and her overall annual job evaluation rating was "Fair." Specific comments that contributed to this rating were "late for work," "does not accept criticism well," and "scowls when asked to retype assignments." Ms. James has come into the personnel office to request a job change.

1. **Personnel clerk:** I see, Ms. James, that you haven't been doing too well in the marketing department. What's going on down there?
   Ms. James: Mr. Barber is very difficult to work for. He picks on everything I do and I just know he doesn't like me.

2. **Personnel clerk:** Well, Mr. Barber feels that *your* attitude is poor and that you are not doing your work as quickly or as accurately as he needs it.
   Ms. James: I'm doing it as well as I can. I'm much faster than the other girls in the office and he never picks on them. Why doesn't he leave me alone? He always hovers over my desk. *(shudders)*

3. **Personnel clerk:** Look, Ms. James, we both know that jobs are tight now and that you need to work. With such a poor reference from Mr. Barber, I couldn't possibly place you anywhere else in this company. Try real hard in the next couple of months to do better and then, after your 18-month evaluation, we'll see what we can do.
   Ms. James: *(sighing)* O.K. I guess I'm stuck. I still don't think I can ever make him like me, though.

How would you rate each of the personnel clerk's (helper's) statements? (Write a number from the scale in the answer blank for each statement.)

0	1	2	3	4	5
Not helpful		Moderately helpful			Helpful

1. ___
2. ___
3. ___

What were the reasons for your ratings? What verbal behaviors from Tables 2.1 and 2.2 can you identify? What would you have said if you were the personnel clerk? How do you think Ms. James was feeling during this session? What was the personnel clerk doing? What do you think the problem really was—and whose problem was it? These are the kinds of questions we continually ponder when participating in or observing helping relationships. Now look at my comments at the end of the chapter.

Exercise 2.5   Ms. Smith is an after-school day-care worker at the local elementary school. One day, while she is supervising the children during their snack break, Tommy runs up to her and angrily denounces Steven for stealing his yogurt. In reviewing the situation, Ms. Smith discovers that Steven has indeed taken and consumed Tommy's yogurt.

1. **Ms. Smith:** Steven, why did you take Tommy's yogurt?
   Steven: I dunno.
2. **Ms. Smith:** Don't you know that it's wrong to steal?

Steven: Uh-huh.

3. **Ms. Smith:** I'm going to have to tell Mr. Singer about this. You go sit in the corner until he comes back.

Steven: O.K.

How would you rate each of Ms. Smith's statements?

0	1	2	3	4	5
Not helpful		Moderately helpful			Helpful

1. __
2. __
3. __

Again, try to determine the reasons for your ratings. What verbal behaviors from Tables 2.1 and 2.2 can you identify? How might you have handled this situation? How do you think Steven feels? How does Ms. Smith feel? What is she doing?

Exercise 2.6   Julie, age 22, has come to see her college counselor. She has been married for two years and is seriously contemplating divorce. (This information was given over the telephone to the counselor when the appointment was made.) The following is an excerpt from the middle part of the first session.

1. **Counselor:** Can you tell me about how things are in your marriage now?

    Julie: Pretty bad. He's always working—we never do anything together. We really have nothing to say to each other. I'm so bored I could scream!

2. **Counselor:** Things are pretty tough for you.

    Julie: *(beginning to cry)* He never wants to go anywhere or do anything. He doesn't like to be with people. We don't have any friends together at all.

3. **Counselor:** You feel angry, as if you can't go out and do what you want on your own.

    Julie: Yes. But sometimes I do go out. Like last Sunday I asked Steve if he wanted to go to the ballet with me. I asked him three times. He said no, I should go by myself. I did, too!

4. **Counselor:** Yet you felt bad about leaving him.

    Julie: I don't want to hurt him.

5. **Counselor:** You feel as if you are responsible for him.

    Julie: Well, yes, of course . . . everything I do . . . . I worry about what's going to happen to him.

6. **Counselor:** Wow! That's quite a burden of responsibility you're carrying around on your shoulders.

    Julie: What do you mean?

7. **Counselor:** I was just thinking that it's hard enough to be responsible for one's self. You can be responsible *to* someone else, but if you take on

the responsibility *for* all of his or her feelings, thoughts, and actions, that's quite a load! It would scare me.

Julie: Yeah, I see what you mean. But aren't I responsible for my husband's feelings?

8. **Counselor:** You tell me.

Julie: Maybe that's why I always feel so hemmed in. I never seem to have any fun.

9. **Counselor:** Tell me about the last time you had fun.

Julie: Last summer I worked as a waitress in a restaurant. It was great. I loved it. I felt so alive and I was with people all the time. I really loved it.

10. **Counselor:** You seem to miss that a lot.

Julie: I had to stop to go back to school. I didn't want to stop though.

11. **Counselor:** You think a lot about going back to that kind of job.

Julie: Yes, I wish I could. I really liked all the attention I got.

12. **Counselor:** It was exciting to have people notice you and respond to you.

Julie: Um-mm. I don't want to miss it all.

13. **Counselor:** Miss?

Julie: You know, the fun and the excitement . . .

14. **Counselor:** You feel as if you don't have any fun and excitement with Steve.

Julie: That's right. Not for a long time, probably never.

15. **Counselor:** You're really angry at Steve. You feel cheated, as if he is keeping you from having any fun and excitement.

Julie: Yes, that's it. Please, can you help me?

Although this was a rather lengthy excerpt, let's see if you can go through the same process as before and rate each response according to the scale.

0	1	2	3	4	5
Not helpful		Moderately helpful			Helpful

1. __   9. __
2. __   10. __
3. __   11. __
4. __   12. __
5. __   13. __
6. __   14. __
7. __   15. __
8. __

You will be able to identify several different verbal behaviors and feelings of both the counselor and the client in this lengthy transcript. What do you think the counselor is doing and why? See if you can trace Julie's feelings through the excerpt.

Exercise 2.7   Joaquin, a 16-year-old high school sophomore, is called to his counselor's office. He had received three warnings in academic classes and is in danger of being removed from the school soccer team. Joaquin's mother has called the counselor several times. She wants him to talk with Joaquin and make him see how important it is for him to do well in school.

1. **Counselor:** Joaquin, your mother is very concerned about your grades and I'm wondering what's going on with you. You know that you're a smart kid and there really is no reason for you not to be getting at least all B's.

   Joaquin: It's O.K. I'm not going to flunk or anything like that. They're just warnings. My mom just worries too much.

2. **Counselor:** Joaquin, come on now. Your parents have every reason to worry because they love you. They've worked very hard since they came to this country and it means everything to them that you do well so you can go to college. They've been through a lot.

   Joaquin: I know that. But they're always on my back. They just don't understand. I know they love me and want me to go to college. They worry too much. I'm not doing worse than anyone else.

3. **Counselor:** For now, we need to figure out how we can get you through the term. How many hours a day are you studying after soccer practice?

   Joaquin: I dunno. Aw, sometimes I forget my homework in the morning but I always do it. I really don't know what the big deal is.

How would you rate each of the counselor's statements in this excerpt?

0	1	2	3	4	5
Not helpful		Moderately helpful			Helpful

1. __
2. __
3. __

Exercise 2.8   Mr. Williams, age 32, is seeing his employment counselor about some forthcoming job interviews. He has been laid off for three months now and has a wife and two young children to support. He has had several unsuccessful job interviews over the past two months.

Mr. Williams: I'm really uptight. And so's my wife. She gets mad when I come home from these job interviews with no job. "Who's gonna pay the bills?" she asks.

1. **Counselor:** It really is rough to be in your position. Sounds to me like you're pretty worried about the job interviews we've set up.

   Mr. Williams: *(nodding)* Well, I must be doing something wrong. None of them have even given me a chance. I must be a real dud. Why should these be any different?

2. **Counselor:** You really want to be able to do better and you're worried that maybe you can't.

   Mr. Williams: I don't know. I never used to flub interviews. I get so nervous now.

3. **Counselor:** The tension of being out of work is really affecting everything.

   Mr. Williams: Yes. I think I come on too strong . . . too wound up. You know, it's like I'm trying to please too much. I just can't seem to help it.

4. **Counselor:** Since your next interview is tomorrow, let's take the rest of our time to practice interviewing. It might help you to feel more comfortable.

How would you rate each of the counselor's comments?

0	1	2	3	4	5
Not helpful		Moderately helpful			Helpful

1. ___   3. ___
2. ___   4. ___

☐ Characteristics of Effective Helpers

Now that we've spent some time differentiating between helpful and unhelpful kinds of verbal communication, we can look more specifically at the characteristics of the effective helper. Review the lists of helpful and unhelpful behaviors shown in Tables 2.1 and 2.2, which you may have modified as a result of working through the exercises, and then try to generalize about helpful and unhelpful traits of the effective helper. However, don't be too quick to draw conclusions.

Human relations trainers are aware that students who come into their programs with certain personal attitudes and traits seem to absorb and integrate their academic training with their living style more easily than other students. Yet when we try to identify the traits that support students' progress as helpers, we—both teachers and students—become rather vague. We make reference to "emotional maturity," "flexibility," "open-mindedness," "intelligence," "warmth," and "sensitivity"—yet these subjective terms tell us little that we can use directly to promote professional growth.

Professional training traditionally involves the study of the academic disciplines of psychology, sociology, anthropology, and the specialized knowledge and skill areas of counseling. Yet an increasing amount of evidence supports the idea that helpers are only as effective as they are self-aware and able to use themselves as vehicles of change. Therefore, training in content may not be as important as training in process, skills, and self-knowledge (Corey, 1990; Kottler, 1991; Okun, 1990). Knowledge in academic disciplines and theory is important, but human relations training programs need to focus more on trainees' self-awareness and communication skills. Integration of personal experience with supervised field experience and academic training becomes essential for developing ability as a helper.

What Research Shows

An initial review of the research about helper characteristics can be overwhelming in that we may wonder if "self-actualized" human beings (those who have achieved self-understanding and fulfillment) really exist! However, if we focus on the commonalities in key research studies, we can identify the characteristics of effective helpers.

Combs (1989), reviewing 13 studies about effective and ineffective helping relationships, concluded that there are some shared beliefs among helpers in the major helping professions, such as the following:

1. *Attitude toward other people*: The effective helper views people as being able rather than unable, worthy rather than unworthy, dependable rather than undependable, helpful and friendly rather than hindering and alienating, optimistic about others rather than negative.
2. *Self-concept*: Effective helpers feel personally adequate rather than inadequate, identify with others rather than feel isolated, feel trustworthy rather than untrustworthy, feel wanted rather than unwanted, feel worthy rather than unworthy.
3. *Approaches to helping*: Effective helpers are more directed toward people than things and are more likely to approach helpees subjectively or **phenomenologically**—that is, from the helpee's vantage point and perspective rather than from their own. The strategies that they use are implemented empathically and are congruent with their own values.

Rogers (1958) believes that helpers must be open and that the following conditions are necessary for helpee development in a helping relationship:

1. *Unconditional positive regard*: Helpers should communicate acceptance of helpees as worthwhile persons, regardless of who they are or what they say or do.
2. *Genuineness and congruence*: Helpers should be real and sincere, honest and clear. They should speak and act congruently in the helping relationship. In other words, they should practice what they preach.
3. *Empathy*: Helpers should be able to communicate empathic understanding of helpees' frames of reference and should let them know they feel and understand helpees' concerns from the helpees' points of view.

Developing Rogers's theoretical formulations into applied research, Carkhuff and Berenson (1967) identified four basic traits that facilitate effective helping relationships if communicated skillfully. They demonstrated that paraprofessionals as well as professionals can receive training to develop those traits and communicate them more effectively and that this increased communicative ability has a positive effect on the helpee's development and change. The traits are as follows:

1. *Empathy*: Effective helpers are able to communicate to the helpee their own self-awareness and understanding, providing the helpee with an experiential base for change.
2. *Respect and positive regard*: Effective helpers can communicate warmth and caring.
3. *Genuineness*: Effective helpers are honest with themselves and their helpees.
4. *Concreteness*: Effective helpers respond accurately, clearly, specifically, and immediately to clients.

Corey (1990) implores helpers to learn about themselves as persons in the helping relationship. He stresses the following conditions:

1. *Self-awareness*: Helpers should continuously develop their awareness of their own values and feelings in order to grow, be open to change, and model congruent behavior and high-risk activity.
2. *Interest*: Helpers should show interest in and involvement with the welfare of others and the influence of culture on all people.
3. *Knowledge and skills*: In order to be professionally effective, practitioners need to be able to integrate psychological theory and practice into their personal meaning.

Egan (1990) refines the concept of empathy by defining two types: *primary empathy*, in which helpers attend, listen, and reflect to communicate accurate perception of the client's message; and *advanced accurate empathy*, in which, in addition to communicating primary empathy, the helper influences the helpee through self-disclosure, directives, or interpretation.

Brammer (1981; Brammer, Shostrum, & Abrego, 1989) believes that the following conditions are necessary to the effective helping relationship:

1. *Self-awareness*: Helpers should be aware of their own values and feelings, of the use (and power) of their ability to function as models for helpees.
2. *Interest*: Helpers should show interest in and involvement with people and social change.
3. *Ethical behavior*: Helpers should demonstrate commitment to behaviors that are reflections of their own moral standards, of society's codes, and of the norms of the helping profession.

Ivey, Ivey, and Simek-Downing (1987) have summarized the findings we've just discussed into what they consider to be the qualitative communication components necessary for effective helping.

1. *Empathy* (primary and advanced accurate): Defined as Egan defined it.

2. *Positive regard*: Drawing on positive assets of the helpee by selectively attending to positive aspects of his or her verbalization and behavior.
3. *Respect*: Stating positive opinions of the helpee and openly and honestly acknowledging, appreciating, and tolerating differences.
4. *Warmth*: Showing concern for the helpee through nonverbal expression.
5. *Concreteness*: Clarifying facts and feelings specifically.
6. *Immediacy*: Speaking in the present instead of the past or future tense.
7. *Confrontation*: Discussing differences, mixed messages, incongruities, and discrepancies between verbal and nonverbal behaviors.
8. *Genuineness*: Being authentic, spontaneous, and sensitive to the needs of the helpee.

In more recent work, Ivey (1991) has proposed a developmental counseling and therapy model where the goal is for helpers to be able to function in four cognitive dimensions at different levels of the helping relationship. This enables them to match their helping style and interventions to the developmental level of the helpee. Ivey is particularly excited about the implications of this neo-Piagetian constructivist model for cross-cultural relevance as he attempts to relate the cognitive level, developmental challenges, and ethnic and gender identity. The four cognitive dimensions in which helpers should be able to function are as follows:

1. *Sensorimotor/elemental*: Helpers should be able to work with clients who are experiencing random, disorganized sensations and who become absorbed in their own or others' emotions. They should be able to ask questions about helpees' feelings and body sensations and to use Gestalt strategies.
2. *Concrete/operational*: Helpers should be able to give and get from clients specific, concrete examples and descriptions of their experiences without much reflection or analysis. They should be able to ask for specific details of what happens before, during, and after an event, what the helpee does, and so forth, and to use behavioral strategies.
3. *Formal/operational*: Helpers should be able to discern patterns of thought, emotion, and action, to abstract, and to help helpees do likewise. Using responsive listening to help clients talk more abstractly about themselves and others' perspectives, they should be able to use cognitive/social learning strategies.
4. *Systemic*: Helpers should be able to integrate and transform patterns of thinking, feeling, and behaving so that helpees can consider other perspectives. They should be able to encourage helpees to look at the role of the family system and larger sociocultural systems in their values, attitudes, and problems.

We can see from the research findings we've covered that the characteristics of the effective helper have much in common with the supportive communication behaviors listed in the previous section. It is hard to separate the helper's personality characteristics from his or her levels and styles of functioning. Likewise, it is difficult to assess whether the knowledge base of understanding behavior or the ability to communicate that understanding has greater influence on a helping relationship.

Characteristics in Context

Now let's see how the characteristics of the effective helper work in context. I believe that the following qualities, behaviors, and knowledge of helpers are most influential in affecting the behaviors, attitudes, and feelings of helpees. And those qualities, behaviors, and knowledge are the same for professional, paraprofessional, and nonprofessional helpers in any context.

Self-Awareness

Individuals who continually develop their own self-understanding and self-awareness are more likely to be effective as helpers than those who do not, because they are more able to separate their needs, perceptions, and feelings from those of their clients and are more able to help others develop their own self-awareness. Development of self-awareness also allows helpers personally to experience the process, both pleasant and painful, of human development and the impact of societal, cultural, and familial influences on behavior. Self-awareness can result in more effective use of the self as a vehicle to effect change in the helpee. Helpers who are self-aware continually ask themselves questions such as: What's really going on here? How come I'm feeling this way? Am I really listening to what is being said or am I projecting my own perceptions and feelings? Whose problem is this—mine or the helpee's?

The following is one example of how self-awareness plays a role in counseling.
A counseling intern in an elementary school is working with a sixth-grade student named Elizabeth. Elizabeth is telling the counseling intern how upset she is with her mother because her mother will not let her drop out of Girl Scouts and participate instead in after-school sports.

> **Elizabeth:** I really want to stay for after-school sports, but my mom just won't let me. She says I've been in Girl Scouts for so long that it would be a shame to drop it now.
> **Counselor:** You are angry with your mother for not letting you do what you want to do.
> **Elizabeth:** She's always on my back, always telling me what I can and can't do. I just wish she'd leave me alone.
> **Counselor:** I don't blame you for being so angry, Elizabeth. Why don't you just tell her that you're old enough to make your own decisions and that you want her to leave you alone?

Elizabeth: Do you really think I can do that? Boy, she might really cream me.
Counselor: She has to learn to let you lead your own life.

Remember that this client is an 11-year-old girl! Although she may be experiencing a normal developmental desire for more independence, should she lead her own life? Instead of helping Elizabeth to explore and understand her independence/dependence conflict, the counselor is giving her permission to rebel. A discussion of this excerpt with the counselor made it apparent that the counselor was overidentifying with Elizabeth and putting words into her mouth that she, the counselor, would like to be able to say to her mother. A more self-aware counselor might say to herself, "This sounds familiar to me. I can certainly understand Elizabeth's feelings, but I must be careful not to let my own needs and feelings interfere with hers."

Gender and Cultural Awareness

Helpers who are sensitive to the influence of gender and culture on their own perceptions, values, attitudes, and beliefs are likely to be open to the effects of these variables on others. For example, they can appreciate that women and minorities have been socialized in nondominant cultures and, therefore, have necessarily experienced their life differently from white males who have been born with more privileges and may never have experienced any forms of oppression. They are likely to be more able to understand and feel comfortable with differences due to gender, race, and ethnicity between themselves and others and to value rather than denigrate these differences.

In other words, they can think and function in what Ivey (1991) terms the systemic cognitive dimension. They recognize the strong set of Western assumptions underlying helping theories and techniques, and realize that helpees from non-Western cultures may have altogether different perceptions of their problems and what to do about them. Thus, they know that they may need to modify and adapt traditional communication and counseling skills when working with helpees from diverse cultures.

Consider the following example.
A job placement counselor describes a night-shift technical job to a recent Russian immigrant, a 34-year-old woman named Anya. The client listens politely, and, through an interpreter, explains why she can't consider night work.

Anya: I no can work night. My mother needs me. I must take care my mother.
Counselor: You're feeling responsible for your mother and as much as you need a job, you're not sure how you can do this. It must be difficult for you, being in a new country with strange ways and needing to find a job but still meet your family obligations.
Anya: I need work, yes. I need money. I need to be good daughter.

Counselor: I admire your sense of family loyalty, Anya. But I'm concerned about how we're going to be able to find a job for you. It's been over a month since you first came in and this is the first time a job for which you are qualified has come up. You must realize this. I don't know if or when another opportunity will arise.

Anya: I need work . . . I need take care mother . . . I need make life.

While the counselor is somewhat sensitive to Anya's dilemma, he is operating on the American premise that finding a job must be the first priority. He would be more helpful to Anya if he were more aware of the issues of immigration, the Russian cultural expectations that adult children will take care of their parents, and the slow process of acculturation. In other words, rather than covertly pressuring Anya to take this job, he needs to explore her confusion and conflicts with her. He needs to know what kinds of questions to ask her to facilitate her seeking resources in the Russian-speaking community to help with her mother and aid her in adjusting to her new life.

Honesty

One of the major variables in developing trust, honesty is a crucial ingredient for any effective interpersonal relationship. We may not always agree with what someone says, but if we believe the other person is being honest, we can respect that person. Helpers can communicate honesty by being open with clients, by answering questions to the best of their ability, and by admitting mistakes or lack of knowledge.

Honesty is more than just being truthful; it is also being open to exploration and being fair in evaluation. One way to assess your own honesty as a helper is to invite honest feedback from clients and peers to see how they view you.

Consider the following example of honesty in a helping relationship.

A registered nurse who has had specialized training in rape counseling is on call at a city hospital emergency ward that provides follow-up counseling to rape victims. She is talking with Mary, a 23-year-old woman who was assaulted three days earlier.

Counselor: You're feeling a lot of pressure from your family and the police to press charges against this guy, aren't you?

Mary: It's been just awful and I don't know what to do. Everyone's been telling me what I should do, what the right thing is. Please tell me—you've been involved with this kind of business before. What do you think I should do?

Counselor: It's quite a dilemma and one that only you can resolve. There really is no one right answer for everyone, you know.

Mary: What would you do if you were me?

Counselor: I honestly don't know, and I've thought about that a lot. I'd like to think I'd have the guts to testify, but after what I've seen, I'm not so sure.

It's very rough going, takes a lot of time, and there's continual harassment that you really have to be tough to take. Not many people can do it, and it doesn't mean that you're not brave or good if you can't.

The counselor was then able to provide factual information about the judicial process in these types of cases and help Mary explore and understand her feelings and thoughts about the information. Because the counselor was open about her own views on the matter, she created a climate of trust and openness. At one point later in the session, Mary expressed appreciation of the counselor's "no-nonsense, honest approach."

You may be wondering how you can be honest in a situation in which you find yourself unable to like or agree with the helpee. These situations do occur, and when they do, helpers need to admit their negative reactions and to separate them from their dealings with the helpee. Occasionally helpers will have to refer a helpee to another helper. This topic will be further explored in Chapter 9.

Congruence

Today's youth commonly accuse the adult generation of hypocrisy—that is, of incongruence or inconsistency between their words and actions. Perhaps incongruence occurs when people have not engaged themselves in a conscious process of examining, clarifying, and acknowledging their values and beliefs.

An individual who experiences congruence between his or her values and beliefs and his or her lifestyle communicates more credibility and has a greater effect as a model than one whose energy is used to deny incongruence. Further, if one has clarified and "owned" one's value system, one is better able to express one's values and beliefs without imposing them on others, thus allowing a more honest, nonjudgmental relationship.

If you believe that the purpose of a helping relationship is to facilitate the helpee's self-understanding and decision making and not to impose the helper's standards and values on the helpee, you'll agree that congruence, which in turn depends on the helper's self-awareness, is an important factor in effective helping relationships. This point of view does not imply that there is a "right" or "wrong" value system for helpers. Rather it simply advocates congruence among what we believe, what we say, and what we live. Furthermore, it seems that people who are aware of and secure in their own values and beliefs are not threatened when faced with divergent or contrasting values and beliefs. Therefore, such people are better able effectively to help a broad spectrum of people.

An example of congruence is seen in the following exchange.

A welfare worker is making a home visit to the Becker family. The Beckers have six children, live in a two-room apartment, and have been on welfare since

Mr. Becker lost his job a year ago. They are having a very difficult time budgeting for sufficient food.

> **Welfare worker:** I don't know how anyone can survive a visit to the market these days, food stamps or not!
>
> **Mrs. Becker:** I just can't manage, that's all. We're lucky if we have meat once a week, the way things are. There just isn't enough.
>
> **Welfare worker:** It's frustrating and also challenging. I do believe we can all find creative ways to change our eating habits for the better.
>
> **Mrs. Becker:** It's impossible, believe me. Easy for people like you to say. You try to feed a family of eight on this kind of money . . .
>
> **Welfare worker:** I'm not saying it's easy, but it is being done. Let me tell you what I've been doing—collecting new casserole and leftover recipes and now we're having more fish, cheese, and poultry and much less beef. Several of us in the office have started to collect a pool of new recipes from ourselves and our clients. I'd be glad to send you some and maybe you could give us some of yours to spread around.
>
> **Mrs. Becker:** No fooling! You people are really doing that?

This incident actually did occur, and is an example of a helper implementing with action what she says (practicing what she preaches).

Ability to Communicate

As discussed earlier in this chapter, the ability to communicate verbally and nonverbally what we perceive, feel, and believe is an aid to any interpersonal relationship. Research substantiates that developing and using communication skills have a positive effect on helping relationships. Let's carry this a step further by suggesting that we can effect positive human relations by continuously teaching (instructionally and by modeling) to our clients and others in our lives the same communication skills that we attempt to master formally or informally. (Examples of communication skills are given in Chapter 3.)

Knowledge

As you will see in Chapter 5, a knowledge of the theories on which effective helping is based is essential to the professional practitioner. Professional helpers need knowledge in the areas of both psychological and sociological theory. Within psychological theory, they need to know about normal and abnormal development, learning and motivation, personality and gender development, and systems theory. Within sociological theory, they need to know about roles, organizations, and cross-cultural influences on behavior.

The formal study of cross-cultural and gender variables in psychological and sociological theory is a recent addition to academic institutions. As well as needing some knowledge of the different cultural backgrounds helpees possess, professional helpers need to consider (1) the usefulness of mainstream psychological theories and practice with different genders and cultures, and (2) the experiences and status of helpees within their own culture in comparison with their experiences and status within a dominant culture. There will always be intracultural as well as intercultural differences.

To a lesser degree, this knowledge base is also relevant for the paraprofessional. Likewise, knowledge of research and applied findings makes it possible for the nonprofessional helper to do a more meaningful job. In any case, the more knowledge one has about social, political, economic, cultural, and psychological issues, the more helpful one can be in aiding people to increase their self-understanding and effective decision making.

How disastrous a lack of some theoretical knowledge of development can be is illustrated in the following situation.

Mr. Howard, a sixth-grade teacher, referred two of his female students to a consulting psychiatrist because he was concerned about their "homosexual tendencies." When asked to share his observations, Mr. Howard said that these two girls held hands, wrote notes to each other during class, always went to the girls' room together, and were really inseparable. Although Mr. Howard was a counselor-in-training, he claimed that his role as a teacher prohibited him from talking to the girls directly about his concerns.

The consulting psychiatrist was quick to point out that there was no basis whatsoever for Mr. Howard's conclusions about these two girls. A basic knowledge of developmental psychology (as well as an awareness of his personal values) would have assured Mr. Howard that same-sex peer affiliation is indeed normal at preadolescent ages—that the need for close relationships is prevalent at this age.

One further comment on counselor effectiveness: there is little doubt that experience enhances a helper's expertise. Research illustrates that increased experience leads to greater helper adaptability to clients and to the use of more eclectic strategies (Corey, 1990; Hutchins, 1984; Kottler, 1991). With experience, helpers learn to trust their intuition, to work with a broader array of helpees, and to risk new approaches.

Ethical Integrity

Overlapping the characteristics of honesty and congruence is a conscious determination on the part of the helper to behave responsibly, morally, and ethically. Deciding what helping behavior is responsible, moral, and ethical involves continued reflection, understanding of potential conflicts among personal, societal, and professional ethical criteria, and openness to change. Ethical dilemmas are complex and challenging. There are no simple solutions. Helpers need the capacity to tolerate ambiguity, uncertainty, and ambivalence. These states can be quite uncomfortable as well as risky. Perhaps an overriding requirement is the ability to value others' needs over one's own when appropriate.

Recently, a colleague of mine who worked in a counseling agency was informed by the director of the agency that she had to terminate immediately those clients who subscribed to a particular insurance plan. While my colleague had informed her employer upon hiring that she was not yet licensed to receive third party payments, he had unwittingly assigned her clients requiring reimbursement. When the claims for these patients were rejected, the director panicked and insisted upon sudden termination. My colleague was eager for work but disturbed by this directive. After terminating two clients and recognizing how upset they were at ending the counseling just when they were getting into it, she began to question the ethics of this behavior. As she pondered the dilemma and consulted other professionals, she decided that sudden termination violated her personal and professional code of ethics, and she confronted her employer. As a result, she lost her job. Although she was distressed by this outcome, she felt comfortable with her ethical stand and was able to insist upon conducting her remaining terminations according to her own values over several sessions, instead of abruptly and unexpectedly.

☐ Helper Self-Assessment

To avoid a dependent relationship with a helpee, it's important to be aware of your own needs, feelings, and problems. As discussed previously, self-awareness enables you to communicate better your equality with, respect for, and confidence in the helpee and your ability to understand empathically the helpee's problems without adding to them by projecting your own feelings and needs. Unfortunately, helpees are not always quick to recognize relationships in which the helper becomes dependent on the helpee's dependence or neediness, perhaps because helpees are often looking for someone to take over for them and tell them what to do. Part of your responsibility is to refuse to do that and to encourage helpees to take responsibility for themselves.

It's useful for helpers continually to assess their own needs and feelings—to think about where they are at any particular time by asking themselves the following kinds of questions and by discussing them with peers and supervisors.

1. *Am I aware of when I am feeling uncomfortable with a client or with a particular subject area?* Very often helpers feel uneasy with a certain type of client who represents something threatening or with a controversial subject such as sex or drugs. It is important for helpers to recognize their discomfort, to "own" it for themselves, and to decide on an honest approach (deal with discomfort and proceed) or avoidance (refer client to another helper). "I" messages and statements can be helpful here—for example, "Look, I find I don't know enough about drugs to really discuss this with you."

2. *Am I aware of my own avoidance strategies?* Do you recognize when you avoid certain topics, allow the client to wander off, or ask too many questions to cover up your insecurities? Helpers who are aware of their avoidance behavior can say to themselves, "This really seems to be bothering me, and I'd better figure out what is going on so I can be truly facilitative with this helpee." One can, therefore, risk failure.

3. *Can I really be honest with the helpee?* Is your fear of being disliked by the client making you afraid to confront or help him or her focus on something unpleasant? Do you have to be perfect and right all the time, or can you be you? If helpers have a strong need to be liked all the time, they will use reassuring, supportive responses to excess and diminish the possibilities for the client's development of responsibility and independence.

4. *Do I always feel as though I need to be in control of situations?* You may have some need for structure and direction in order to be accountable and achieve goals and objectives, but you should be aware of how you feel when a helpee disagrees or wants to pursue something different. For example, there may be times when you want to try a different approach, such as a Gestalt exercise, with the helpee, and he or she refuses to participate. If you have a need to control, you may feel angry and rebuffed in this situation. If you do not have this need to control, you can accept the helpee's feelings without feeling personally attacked and can propose alternatives or delay introducing another approach. Responsive listening is a safeguard against controlling the communication process.

5. *Do I often feel as though I must be omnipotent, that I must do something to make the helpee "get better" so I can be successful?* If you often experience this feeling, you may ask whether you're in the right field! It's your relationship with helpees that will facilitate their resolving their problems to their satisfaction, not the waving of your magic wand. You can feel good about yourself when you see helpees acting for and taking responsibility for themselves.

6. *Am I so problem-oriented that I'm always looking for the negative, for a problem, and never responding to the positive, to the good?* This is a common concern of helpers, since they're exposed more often to negative feelings than to positive ones. However, it's important to identify and respond to positive affective and cognitive content in order to balance perspective and, more important, to reinforce positive conditions.

7. *Am I able to be as open with clients as I want them to be with me?* A common problem of people in helping professions is that they want to avoid their own feelings and problems by focusing on those of their clients. A good rule of thumb is never to ask anyone to do or talk about anything that you would not be willing to do or talk about in that or a similar situation.

Some of the preceding questions deal directly with communication, while others are more closely related to the issues discussed in Chapter 9, "Issues Affecting Helping." Your ability to communicate effectively is inseparable from your continuing development of sensitivity and self-awareness.

☐ Client Variables

We will note when we study the various major theoretical approaches that certain approaches require certain client characteristics. For example, the client-centered and psychoanalytic approaches require a high degree of verbal ability on the part of the helpee, whereas the behavioral approaches can be used with less verbal clients. All approaches require some level of motivation and degree of cooperation on the part of the helpee to participate in the helping process. Some place more responsibility on the helper to develop that motivation, but all assume that the helping relationship will enhance the helpee's willingness to open up and accept vulnerability in order to achieve growth in the affective, cognitive, and behavioral domains.

☐ Summary

A helping relationship involves verbal and nonverbal communication between helper and helpee. Communication facilitates the development of rapport between them, which in turn allows for exploration of the helpee's beliefs, values, attitudes, feelings, and behaviors. The aim is to increase the helpee's self-understanding and understanding of others. The self-aware helpee will possess higher self-esteem, resulting in greater tolerance and acceptance of others. He or she will be better able to decide on and adopt a course of action to attain agreed-upon objectives, and to assume responsibility for the consequences of that action.

There are many different kinds and levels of helping relationships and many different approaches to helping people, but clear communication is the basis of all helping. In informal helping relationships, there is a reciprocity that does not exist in formal helping relationships: helpers and helpees can switch roles in the former.

Experience and research both indicate that certain traits and characteristics of helpers appear to positively affect helping relationships. The more in touch people are with their own gender and culture biases, beliefs, behaviors, and feelings and the more able they are to communicate genuinely, clearly, and empathically their understanding of themselves and others, the more likely they are to be effective helpers. Self-awareness, honesty, congruence, the ability to communicate, and knowledge of human behaviors and the impact of gender, culture, and social factors on behavior all enhance the helping relationship.

Certain communication behaviors—verbal and nonverbal—also affect the process and outcome of helping relationships. The exercises in this chapter should have helped you to recognize verbal and nonverbal behaviors that are consistent with your personal style and beliefs and to become aware of which behaviors different helpees find helpful.

☐ Comments on Exercises

Exercise 2.4   We can't tell from the excerpt who has the problem, Ms. James or Mr. Barber. The personnel clerk assumed that the problem was Ms. James's and made no attempt to verify that assumption by gathering new data from her. By learning more about Ms. James's thoughts, feelings, and behaviors, the personnel clerk might have begun to help her clarify the nature of the problem, what part of it she was responsible for, and what her options and alternatives were for coping with the problem. The personnel clerk should have helped Ms. James to feel worthwhile as a human being, regardless of what she was or was not doing, so that she could retain her dignity; at the same time he or she might have encouraged Ms. James to explore more fully and understand what was happening to her, to take responsibility for herself. The personnel clerk's three statements were not helpful; no clarification or change occurred. Behaviors included judging, blaming, and telling the client what to do. Ms. James felt defensive, as one is likely to feel when attacked. She had come to the personnel office to see about a job change, and she never really got the chance to do so; she was immediately pushed into a corner.

Exercise 2.5   This short excerpt is all too typical of the interactions between adults who see themselves as being helpful and youngsters who have been "caught in the act" of doing something wrong. Ms. Smith was not helpful, since she did not help Steven to retain his feelings of self-worth and dignity or to understand his own behavior. Nor was Steven helped even to begin to verbalize what the problem really was all about, much less to understand what he could do about such problems in the future. Ms. Smith was apparently not concerned with Steven's feelings, nor was she concerned with trying to understand him. Behaviors that she used are judging, punishing, threatening, and blaming. She also asked a "why" question, which ensures defensiveness. If people knew why they did things that cause trouble, they would not have to get into compromising situations. Questions such as those of Ms. Smith cut off communication and encourage denial and withdrawal. Exploration and understanding cannot occur if communication is cut off. Like Ms. James, Ms. Smith most likely felt angry and frustrated at not being able to control a situation. A more appropriate response for Ms. Smith would have been "Steven, it's very difficult when somebody else has something that you really want." This response would have allowed Steven the opportunity to tell his side of the story to an empathic listener.

Exercise 2.6   Julie's counselor demonstrated effective responsive listening skills and succeeded in eliciting much affect (feeling) and **content** (information) from her. Some of the responses were more helpful than others, but overall this was a helpful first session in what subsequently became a short-term helping relationship. (At the end of this helping relationship, Julie was able to decide for herself and take responsibility for her decision and action regarding her marriage.)

Behaviors that the counselor used were reflection, empathy, clarification, and interpretation. This is a lengthy excerpt, so let's examine each of the counselor's statements.

1. This is a good opener, eliciting more information on an open-ended level.
2. This is a reflective, empathic statement in that it conveys the counselor's understanding of Julie's intense feeling.
3. Again, this statement clarifies the problem by focusing on Julie's feelings.
4. Some interpretation is involved here in that the counselor interprets the nonverbal behavior that accompanied this statement (nonverbal behavior was apparent from the tape of this case).
5. Again, there is some interpretation. The client's response indicates that the counselor was on target.
6–7. Counselor focuses on the issue of responsibility, adding some possible materials for client to consider.
8. Counselor backs away from imposing own values here and succeeds in putting the issue back in the client's hands.
9. Another exploratory statement—open-ended.
10. There is some interpretation of the underlying message here.
11. This is a probing statement. The counselor is not quite sure if this is the right track.
12. Again, probing.
13. Strictly reflection.
14. This response reflects the client's statement but brings the conversation back to her relationship with her husband.
15. The counselor is responding to the intense underlying feeling and focusing this feeling on the husband.

In this excerpt, the counselor had to make some decisions about what she would focus on with the client; there were many different directions she could have followed, and one direction was not necessarily more right than another. Julie was helped to begin to explore what was bothering her and to begin to understand some of her feelings. From this beginning the counselor could continue to develop the relationship and help Julie assume responsibility for herself and make her own decisions.

Exercise 2.7 Joaquin's counselor did not even recognize Joaquin's feelings, much less show concern for them. He totally ignored the possibility of Joaquin's underlying yearnings for assimilation and assumed that Joaquin's mother's concern was stereotypical immigrant upward-mobility anxiety. Thus, he was not helpful to Joaquin at all because he did not explore Joaquin's feelings and he imposed his own needs on Joaquin by focusing exclusively on "getting through the term." If Joaquin had been allowed to explore his feelings and his desire to become more independent of his parents, he might have decided to put more effort into his schoolwork. Counselor behaviors used here were projecting, telling the helpee what to do, and judging. The first statement was moderately helpful in that it allowed Joaquin the open-ended opportunity to pursue what was on his mind and communicated positive regard for Joaquin's academic capabilities. However, the counselor's second and third statements were not helpful in that they conveyed judgment and a desire for expedience. Thus, Joaquin did not have the opportunity to learn about himself and his real concerns so as to decide for himself how to function in school.

Exercise 2.8   Mr. Williams's counselor demonstrated helpful communication skills. He took his time in drawing out his client so that it was the client, not the counselor, who clarified the problem by elaborating on his "coming on too strong" in job interviews. A less patient counselor might have simply accompanied the client in bemoaning his misfortune rather than clarifying what might be done about it. Each of the counselor's three responses demonstrated empathy, reflection, and congruence. The counselor's last statement was action-oriented, taking into consideration the nature of the problem and the available time for intervention.

☐ References and Further Reading

Atkinson, E. R., Morten, G., & Sue, D. W. (Eds.). (1989). *Counseling American minorities: A cross-cultural perspective* (3rd ed.). Dubuque, IA: William C. Brown.

Benjamin, A. (1981). *The helping interview* (3rd ed.). Boston: Houghton Mifflin.

Brammer, L. (1981). *The helping relationship* (3rd ed.). Englewood Cliffs, NJ: Prentice-Hall.

Brammer, L., Shostrum, E., & Abrego, P. J. (1989). *Therapeutic psychology: Fundamentals of counseling and psychotherapy* (5th ed.). Englewood Cliffs, NJ: Prentice-Hall.

Carkhuff, R. (1983). *The art of helping* (5th ed.). Amherst, MA: Human Resource Development Press.

Carkhuff, R., & Berenson, B. (1967). *Beyond counseling and therapy*. New York: Holt, Rinehart & Winston.

Combs, A. (1989). *A theory of therapy: Guidelines for counseling practice*. Newbury Park, CA: Sage.

Corey, G. (1990). *Theory and practice of counseling and psychotherapy*. Pacific Grove, CA: Brooks/Cole.

Corey, M., & Corey, G. (1989). *Becoming a helper*. Pacific Grove, CA: Brooks/Cole.

Cormier, L. S., & Hackney, H. (1987). *The professional counselor: A process guide to helping*. Englewood Cliffs, NJ: Prentice-Hall.

Egan, G. (1990). *The skilled helper* (4th ed.). Pacific Grove, CA: Brooks/Cole.

Hutchins, D. E. (1984). Improving the counseling relationship. *Personnel and Guidance Journal, 62,* 572–576.

Ivey, A. E. (1991). *Developmental strategies for helpers: Individual, family, and network interventions*. Pacific Grove, CA: Brooks/Cole.

Ivey, A., Ivey, M., & Simek-Downing, L. (1987). *Counseling and psychotherapy: Integrating skills, theories, and practice*. Englewood Cliffs, NJ: Prentice-Hall.

Kottler, J. A. (1991). *The compleat therapist.* San Francisco: Jossey-Bass.

Okun, B. F. (1989). Therapists' blind spots related to gender socialization. In D. Kantor & B. F. Okun, *Intimate environments: Sex, intimacy, and gender in families* (pp. 129–163). New York: Guilford.

Okun, B. F. (1990). *Seeking connections in psychotherapy.* San Francisco: Jossey-Bass.

Pedersen, P. (1988). *A handbook for developing multicultural awareness.* Alexandria, VA: American Association for Counseling and Development.

Pietrofesa, J., Leonard, G., & Van Hoose, W. (1971). *The authentic helper.* Chicago: Rand McNally.

Rogers, C. (1951). *Client-centered therapy.* Boston: Houghton Mifflin.

Rogers, C. (1958). The characteristics of a helping relationship. *Personnel and Guidance Journal, 37,* 6–16.

Rogers, C. (1975). Empathy: An unappreciated way of being. *The Counseling Psychologist, 5*(2), 2–10.

Strong, S. R. (1968). Counseling: An interpersonal influence process. *Journal of Counseling Psychology, 15,* 215–224.

THREE

Communication Skills

To be effective, helpers must use communication skills that involve hearing verbal messages (cognitive and affective content), perceiving nonverbal messages (affective and behavioral content), and responding verbally and nonverbally to both kinds of messages. To ensure that communication skills become an integral part of your helping techniques, you must practice them frequently. Because you, like most people, have probably taken your communication behaviors for granted, you may not have had the opportunity to focus on and develop an awareness of them. When you start learning communication skills you may find yourself concentrating harder on your communication behaviors (these behaviors were briefly reviewed in Chapter 2). At first this concentration may make you feel uncomfortable and tired. It's relatively easy to talk about communication skills and theoretically understand their importance to effective helping relationships; however, it's much more difficult to put this understanding into practice. Indeed, practicing communication skills is more difficult than completing written exercises about them!

Because the practice of communication skills is so important, I suggest incorporating the exercises in this chapter into your group discussion. Sharing what you have experienced during the exercises, your feelings and your recognition of what you are willing to do and what you tend to shy away from (**approach** and **avoidance reactions**), is the most beneficial aspect of the exercises. More understanding comes from the discussion after an exercise than from the exercise itself. In this kind of discussion, called *processing,* one talks about one's feelings and reactions rather than the actual content of the exercise.

☐ Perceiving Nonverbal Messages

The reason it's so important to understand nonverbal communication is that it is the foundation on which human relationships are built. Some anthropologists believe that more than two-thirds of any communication is transmitted on a nonverbal level. We must interpret patterns of gestures, posture, facial expressions, spatial relations, personal appearance, and cultural characteristics. Thus we, as helpers, should try to develop a conscious awareness of nonverbal manifestations and their various meanings.

Because the perception of nonverbal messages has not been emphasized in our culture, some of the exercises in this chapter may produce frustration and tension as we become aware of our dependence on nonverbal cues for understanding verbal messages. Chapter 2 listed various facilitative and nonfacilitative nonverbal behaviors in helping relationships. The kinds of nonverbal cues that we attend to in a communicative relationship are listed in Table 3.1. We, as helpers, look to see if the nonverbal behavior is consistent with the verbal behavior and if we can pick up clues from it that will help us to identify the affective messages (underlying feelings) we hear. Nonverbal behavior provides us with clues to, not conclusive proofs of, underlying feelings. However, research has proven that nonverbal clues tend to be more reliable than verbal clues!

Exercise 3.1   What do the following gestures mean to you? When you have completed this exercise, compare your answers with those of your classmates and those suggested at the end of the chapter.

1. A man walks into your office, takes off his coat, loosens his tie, and sits down and puts his feet up on a chair.
2. A man walks into your office, sits erect, and clasps his arms across his chest before saying a word.
3. Your client rests her cheek on her hand, strokes her chin, cocks her head slightly to one side, and nods deeply.
4. A woman walks into your office, sits as far away as she can, folds her arms, crosses her legs, tilts the chair backward, and looks over your head.
5. A client refuses to talk and avoids eye contact with you.
6. A client gazes at you and stretches out her hands with the palms up.

Table 3.1 Nonverbal Cues in a Communicative Relationship

Feature	Examples
Body position	Tense, relaxed, leaning toward or away from
Eyes	Teary, open, closed, excessive blinking, twitching
Eye contact	Steady, avoiding, shifty
Body movement	Knee jerks, taps, hand and leg gestures, fidgeting, head nodding, finger pointing, dependence on arms and hands for expressing message, touching
Body posture	Stooped shoulders, slouching, legs crossed, rigid, relaxed
Mouth	Smiling, lip biting, licking lips, tight, loose
Facial expression	Animated, bland, distracting, frowning, puckers, grimaces
Skin	Blushing, rashes, perspiration, paleness
General appearance	Clean, neat, sloppy, well groomed
Voice	Fast, slow, jerky, high pitched, whispering

7. A client quickly covers his mouth with his hand after revealing some sensitive material.

8. You are talking to someone who holds one arm behind her back and clenches her hand tightly while using the other hand to grip her wrist or arm.

9. A woman in your office crosses her legs and moves her foot in a slight kicking motion; at the same time she is drumming her fingers.

10. A person sits forward in his chair, tilting his head and nodding at intervals.

Exercise 3.2   Cut out some pictures of people from a magazine. Remove the captions and ask a partner what he or she believes is being communicated. Then, in groups of four to six, ask other people to respond to the same pictures. Talk about the various responses and see if it is possible to reach a consensus. After you have identified the feelings that are being communicated in several pictures, see if you can establish some patterns in your group's identifications. The purpose of this exercise is to examine different responses to the same nonverbal stimulus. What were the reasons given for a particular identification? In cases of disagreement, what were the major areas of difference? How do people project their own attitudes, values, and beliefs onto their perceptions of subjects' feelings? How can you explain the variety of perceptions?

Exercise 3.3   The purpose of this exercise is to give you the opportunity to identify the feelings underlying another person's nonverbal behavior. In triads or small groups, have one person identify a specific feeling or emotion, tell it to an observer, and then try to communicate it nonverbally to his or her partner or to the other members of the group. After the emotion has been properly identified, someone else should choose a feeling and attempt to communicate it to the group. Process your experiences as you did in Exercise 3.2.

Exercise 3.4   Another way to become adept at giving and identifying nonverbal messages is to play a nonverbal form of "telephone." A group sits in a circle and one person selects a feeling to start the game. Everyone closes his or her eyes, and the starter taps the person on the right who opens his or her eyes, tries to understand the nonverbal communication, and then repeats the communication process by tapping the person on the right, who opens his or her eyes and receives and sends the nonverbal message. This entire exercise is nonverbal until the last person in the circle receives the message and verbally identifies the feeling. The group then processes what happened, where and how the message became distorted, if it did, and what the members felt about the nonverbal communication.

Exercise 3.5   This exercise helps identify complex nonverbal messages by providing a scene, rather than the acting out of an isolated feeling, to observe. A pair leaves the room and plans a ten-minute role-play scene of a couple talking at the dinner table, a teacher and principal conferring, or whatever seems appropriate (and fun!). They come back into the room and, using nonsense syllables, carry on a dialogue that has no verbal meaning for the observers. The observers are to identify and reconstruct what is occurring strictly by observing nonverbal behavior. During the processing of this exercise, the role players can contribute by sharing their intentions and feelings. The observers usually find that they can understand the meaning of the scene without verbal information.

Exercise 3.6   The purpose of this exercise is to become more aware of the nonverbal behaviors associated with your feelings. List all of the nonverbal behaviors you can for each of the four major emotions: anger, fear, happiness, and sadness. For example, "When I'm mad, I frown, clench my fists, tighten my body, sit back away from people, and feel knots in my stomach." Share your list in small groups and note similarities and differences.

Exercise 3.7   Read the following client statements and list the major feelings in each of the situations.

1. "I'm really up a tree at this point as I have so many bills to pay and Tom isn't working and I don't know what to do. This job doesn't pay enough and I guess I should see if I can get an extra evening job, like waitressing or telephone soliciting. I wanted to go back to school at night this term, but that doesn't seem possible now."
2. "Why should I stay in school when there are so many people with degrees who can't get jobs anyway? I don't feel as if I'm learning anything here anyway, nobody seems to care what happens to anybody, and the classes are so large and impersonal. It's a waste of money."
3. "I couldn't find a parking place today so I missed this morning's briefing session. Somebody ought to do something about this situation. I can't find that memorandum and I'm not sure what to say to Mr. Jones when he calls this afternoon."
4. "Look, I've got a sick kid, we just moved here and don't know anyone. The landlord won't turn on the heat, and I need to find a job to pay the bills. I don't know where to turn first, but I also need to get the rest of my furniture delivered."
5. "Wow, what a time we had! We went swimming every day, bicycling at night, fishing several times, and just had a great time. Oh yes, we sailed the Meyers's boat—did you know they were down there?"

As you go over your answers, see if you listed your own feelings or what you believe the subject is feeling. How can you tell? How did you decide? Discuss your list in small groups.

Exercise 3.8   In pairs (or triads), take turns being helper and helpee (and observer). The helpee describes a real or imagined concern and uses facial expressions, vocal tones, postures, and gestures that are opposite to the feelings he or she is expressing in the verbal message. For example, I may tell you how excited I am about a trip I'm going to make this weekend and slump, frown, and slur my words. What feeling do you perceive? The helper states what he or she perceives the major expressed feeling to be. Then group members discuss their experiences with verbal and nonverbal incongruence. Which cues did you find yourself giving most attention to? What did you make of this?

Exercise 3.9   The purpose of this exercise is to explore gender- and ethnicity-related characteristics of nonverbal behaviors. As each member of the group acts out a concern or event nonverbally, jot down what you think may be gender or ethnically linked. During the next week, observe nonverbal behaviors of as many different people as possible: on buses, in restaurants, waiting rooms, and so on. At your next group meeting, share your findings and see if your group can agree upon gender- or ethnicity-related differences in nonverbal behaviors.

☐ Hearing Verbal Messages

We are all aware of the need to listen to verbal messages, and we can sometimes accurately restate simple verbal messages for their senders in a one-to-one situation. But if we think back to the "telephone" game that we played at childhood parties, we can remember how simple verbal messages became distorted as they passed through several different people.

As difficult as it sometimes is to understand apparently clear-cut verbal messages, it is much more difficult to understand the underlying affective content of verbal messages. One of the reasons is that we tend to respond more to the cognitive content than to the affective content. We fail to recognize the inconsistencies and the underlying dynamics of the "hidden agenda."

Cognitive content comprises the actual facts and words of the message. Affective content may be verbal or nonverbal and comprises feelings, attitudes, and behaviors. Receiving verbal messages really involves understanding both cognitive and affective content and being able to discriminate between them. The cognitive content is usually easier to understand; it is stated. The affective content sometimes differs from the cognitive content and is often less apparent. The difference between hearing only the apparent cognitive content of a verbal message and hearing both the cognitive and underlying affective messages is the difference between being an ineffective and an effective listener.

Your response to a client's statement will depend on your ability to hear and understand what is being said and to uncover the underlying message. Your response will, in turn, influence the direction of the client's next statement. Thus, before you can learn to respond appropriately to a client's statement, you must learn to hear and discriminate between his or her apparent and underlying cognitive and affective messages.

Verbal Cognitive Messages

As previously stated, cognitive messages are easier for us to recognize than affective messages, for our schooling stresses cognitive knowledge. Cognitive messages usually involve talking *about* things, people, or events and may involve one or several simple or complex themes. The client is often more comfortable talking *about* thoughts or behaviors than actually feeling them.

If we find ourselves responding only to the client's cognitive concerns, we never really get down to his or her underlying feelings. For example, a client may come in and talk about trouble with a supervisor. If the helper asks only "What happened?" "What did your supervisor say?" and "What did you say then?" the entire session can pass without uncovering the client's underlying concerns and feelings. The only possible outcome of this type of helper response would be the helper's suggestion that the client make different verbal responses or offer different behaviors to the supervisor. The helper's suggestions may work out but they really don't contribute to the client's understanding and choice of his or her own course of action. The solution of one obvious problem may not touch on the helpee's underlying concerns.

The theme (problem or concern) that the helpee focuses on affects the direction of the ensuing discussion. A helper hopes to choose a theme that will be the most productive in developing the helping relationship. It is difficult to determine which cognitive theme is most important, and there could be several of equal importance, but it is necessary to respond to and focus on one major theme at a time in order to clarify and explore all aspects of the situation.

The objective of the following exercises is to help you identify cognitive themes in communication.

Exercise 3.10   Read the following client statements (they are the same as in Exercise 3.7) and pick out as many different cognitive topics as you can. Then check your answers with mine at the end of the chapter. Remember, list only cognitive content, not affective content (feelings).

1. "I'm really up a tree at this point as I have so many bills to pay and Tom isn't working and I don't know what to do. This job doesn't pay enough and I guess I should see if I can get an extra evening job, like waitressing or telephone soliciting. I wanted to go back to school at night this term, but that doesn't seem possible now."
2. "Why should I stay in school when there are so many people with degrees who can't get jobs anyway? I don't feel as if I'm learning anything here anyway, nobody seems to care what happens to anybody, and the classes are so large and impersonal. It's a waste of money."
3. "I couldn't find a parking place today so I missed this morning's briefing session. Somebody ought to do something about this situation. I can't find

that memorandum and I'm not sure what to say to Mr. Jones when he calls this afternoon."

4. "Look, I've got a sick kid, we just moved here and don't know anyone. The landlord won't turn on the heat, and I need to find a job to pay the bills. I don't know where to turn first, but I also need to get the rest of my furniture delivered."

5. "Wow, what a time we had! We went swimming every day, bicycling at night, fishing several times, and just had a great time. Oh yes, we sailed the Meyers's boat—did you know they were down there?"

Now go over your list of topics and see if you can rank them in order of importance or immediacy of concern for each statement. If you had to respond to only one of the themes in each case, which would it be?

Because the exercise provides the client statements merely in written form, it is obviously difficult to rank objectively the cognitive topics in terms of importance to the client. However, once you have learned to hear the verbal cognitive content, you will be able to use affective content as a clue for establishing priorities. In verbal communication, we tend to respond more often to the most recent verbal theme rather than to the most important one. We need to train ourselves to hear the whole message and discriminate among themes.

A variation of this exercise would be to read the statements aloud to a partner and have him or her verbally recall the cognitive themes.

Exercise 3.11   This exercise has many different forms and names. Its purpose is to develop students' attending skills by having them repeat verbal cognitive messages. It is effective in groups that divide either into pairs (helper, helpee roles) or into triads (helper, helpee, observer roles). One person, the helpee, communicates to the helper statements that last no longer than three minutes and that involve a real or imagined concern. The helper must then restate the helpee's verbal message to the latter's satisfaction. The helpee and the observer (if there is a triad) then process this interaction by evaluating the effectiveness of the helper's restatement of the helpee's verbal cognitive content. Everyone in the pair or triad should have the chance to be the helpee and the helper. The observer can jot down what he or she has heard and compare it to what the helper restates.

After the pairs or triads have completed this exercise, they should recombine into a group and process their experiences by sharing their difficulties and concerns. Group members should then make suggestions for dealing with those difficulties and concerns. Observers should share their observations about the helper's verbal and nonverbal behaviors, such as posture, eye contact, and gestures.

Exercise 3.11 helps us to realize that we spend more energy preparing our responses than listening to what is actually being said to us. It is also a good exercise to show us how we often hear another person's verbal message through our own "filters" (selective perception). We often hear and see what we need or want to hear and see rather than what is actually being communicated.

Exercise 3.12   In this exercise, choose the helper response that best reflects the cognitive content of the helpee's statement. Remember that you are trying to paraphrase the main idea of each statement without changing it.

The following situation occurred in a professional accounting firm. The helpee is a recent college graduate who was hired six months ago and has been more and more frequently absent and tardy. The helper is the partner of the firm responsible for personnel and office management.

1. **Helpee:** I'm not so sure what it is you people are so uptight about here.
   Helper:
   a. You think we make too much of things here.
   b. You're concerned about how we feel about your absences and tardiness.
   c. How do you feel you're doing here?
2. **Helpee:** I graduated at the top of my class and I know this stuff cold.
   Helper:
   a. Sounds like you think you're better than the rest of us.
   b. You have a lot to teach us.
   c. You seem concerned about how well you're doing here and how you're fitting in.
3. **Helpee:** In the past six months, I've moved to a strange city and had a devil of a time finding an apartment and roommate and trying to make ends meet.
   Helper:
   a. You've had a lot of change in your life recently.
   b. Your personal life is really not related to what's happening here.
   c. Being out in the real world is very different from college life, isn't it?
4. **Helpee:** Well, I'm really getting bored doing what I'm doing. My education seems wasted. How soon can I move on to something more in line with my qualifications?
   Helper:
   a. Sounds like you really need to move on quickly.
   b. You're eager to make more money.
   c. It's frustrating for you to feel that you're not working up to your potential.
5. **Helpee:** I look at some of the guys who've been here for years and really haven't gotten very far. I won't let that happen to me.
   Helper:
   a. It's important for you to get ahead quickly.
   b. Young people today have a hard time working their way up. They want everything to happen so quickly.
   c. You're afraid that you may get stuck and not get to where you want to go.
6. **Helpee:** There's just too much junk work that one has to do around here and too much politicking. I want to get on with it.
   Helper:

a. It's hard for you to do the scut work that needs to get done.

b. You're angry that this job didn't turn out the way you wanted it to.

c. It's hard for you to get yourself to do what you don't want to do.

7. **Helpee:** Sometimes I wonder what I'm doing here. I'm sure I could make more money someplace else.

Helper:

a. You're really unsure about whether or not to stick it out here.

b. You're very worried about your financial affairs.

c. It's confusing to you whether money is more important than the job training you're receiving.

8. **Helpee:** Being on time and sticking in the office is hard for me. I'm used to being able to do what I want when I want.

Helper:

a. The structure here really gets to you.

b. You get angry at us for making demands on you so you just don't bother to come in.

c. Sounds as though you're having a conflict—you really want to work and learn, but you're having trouble getting used to the routine and structure here.

Exercise 3.13   This exercise demonstrates the incongruities that can exist between verbal and nonverbal communication in cognitive messages. Divide into your pairs or triads and repeat Exercise 3.11 with one difference: the helper should deliberately attempt to use facial expressions, vocal tones, posture, and gestures that are opposite in meaning to the cognitive messages being sent. For example, if I talked about my concern about finding a legitimate parking place in time to get to a downtown meeting and at the same time smiled, seemed relaxed, and nonverbally expressed casualness, my actions and my words would be incongruent. The point of this exercise is for the helper to experience how such incongruities between verbal and nonverbal behaviors can affect perception of cognitive messages. Helpees will experience the discomfort involved in communicating incongruous messages on a conscious level.

Verbal Affective Messages

Affective messages are communicated to us both verbally and nonverbally, but for the present we will restrict ourselves to verbal communication. Affective messages involve feelings: emotions that may be directly or indirectly expressed. They are much more difficult to communicate than cognitive messages and much more difficult to perceive and hear. Clients are often so much more aware of thoughts than feelings that the helper's responses, clarifying and identifying feelings, come as a surprise to them and uncover a whole new area for exploration and experiencing. By understanding affective messages and in turn responding to them, the helper is communicating not only acceptance of the helpee's emotions but also permission for the helpee to experience and "own" those feelings.

Feelings can be grouped into four major categories: anger, sadness, fear, and happiness. Very often a feeling from one category covers up one from another (for example, sadness sometimes masks anger or vice versa, or anger may mask fear). There are many different words that we can use to identify feelings in these four categories, and it is helpful to select vocabulary that is comfortable for the client. For example, if a teenager is using the current vernacular of his peer group, instead of using the word "angry" when identifying his feelings, you may say "pissed off" as long as you feel comfortable doing this (and do not come across as a phony). Identifying underlying feelings in verbal messages is difficult at first and is related to how comfortable and proficient you are in recognizing and expressing your own feelings. It is crucial that you listen to the client's messages and identify his or her feelings rather than project your own onto the client. Again, this requires continual, repeated practice.

Exercise 3.14   The purpose of this exercise is to clarify your own ways of expressing feelings and emotions. Write down all the words you can think of that express each of the four major categories of emotion: happiness, anger, fear, and sadness. What words do you use to express those feelings? For example, under "happiness" you may write "glad," "groovy," "tip-top," and so forth. When you have completed your list, share it with two others and see if there are many differences.

Exercise 3.15   Now take each of the four major categories of emotion (happiness, anger, fear, and sadness) and list as many verbal behaviors as possible that you are aware of enacting when you feel each emotion. For example, when you're mad, you may swear, use short, clipped sentences or monosyllables, or yell. Share your lists in small groups and learn how different people express the same emotion.

Exercise 3.16   This exercise is called listening for feelings. For each of the following statements, write what you think the person is really feeling. Ask yourself, What is the underlying feeling here?

1. "Two big boys were picking on me when I was coming home from Boy Scouts today."
2. "The doctor told me to come over here and have all these tests. I'll sit over here and wait until you're ready for me."
3. "Poor Lenny! He works so hard and he never gets home for dinner anymore."
4. "I can't wait until final exams are over."
5. "I'm really too busy to take a coffee break now, though I'd love to talk to you."
6. "I can't type that report today. Professor Greene gave me four rush letters to get out by three o'clock, and I still have the exams to type."
7. "Please put down that newspaper. You never talk to me anymore."
8. "I think people are out to get what they can for themselves."

9. "If Jim hadn't been transferred, this project would have gotten off the ground in plenty of time."

10. "Have you heard anything about the new social worker? I'm supposed to see her at three o'clock."

11. "I hear the new office manager is a real clock watcher."

12. "Only two more weeks until vacation!"

13. "Look, Ms. Jones, if you can't get this typing done, I'll have to see if the typing pool can do it."

14. "My husband is out of work, and I don't know how we're going to pay the rent next month."

15. "All children steal at that age, don't you think?"

16. "Young people today really have a lot more sexual freedom than we did in my day!"

17. "John, I want to tell you that, after much careful consideration, I'm stepping down as chairperson so as to have more time for my family and to do my research."

18. "Ms. Green is a lousy teacher. She doesn't know how to explain things."

19. "Please bring the car home by eight o'clock. I don't want you driving in the dark."

20. "Why should I stay in school? I don't know what I want to do. What do you think?"

21. "I hate staff meetings. No one ever gives me a chance to talk."

22. "Coming to see you just doesn't seem to be helping me. We talk about the same stuff over and over and I still don't know what to do."

23. "No one ever picks me to be on his team at school."

24. "Are you going to see me again this week, doctor?"

25. "I'd like to talk to you when you have a minute. Be sure and see me before you leave the office tonight."

After you have completed this exercise, discuss your answers either in small groups or as a large group. Then look at the answers at the end of the chapter. You will note that different people identify different underlying feelings for the same statement and that, in discussing these statements, various projections begin to emerge. For example, one group of students told me that some members felt so threatened by statement 25 that they identified the underlying feeling as anger, whereas other members perceived it as eagerness, hypothesizing a situation where one person in an office wants to invite another person over but doesn't want others in the office to know about it and feel excluded. Try reading aloud, with varying affect and intonation, the statements on which the group disagrees and see what different kinds of reactions you get.

Exercise 3.17   This exercise is similar to Exercise 3.11 (verbal cognitive messages), but in this case you'll identify the feelings rather than the cognitive content. In pairs, with backs to each other to block out nonverbal cues, have the helpee talk for up to three minutes and have the helper identify the feelings that were being communicated. The helpees should not verbally identify their own feelings for the helpers but should make the same types of comments as they did in Exercise 3.11 and allow the helper to identify the underlying feeling. The results of this exercise should be processed in the same manner as in Exercise 3.11.

Exercise 3.18   Have two pairs (or triads) work with each other, with two people identified as helpers and two as helpees (and two as observers). The purpose of this exercise is to show each helper how two different people may express the same feeling with different verbal messages and different nonverbal cues. The two helpees are to separate themselves from the rest of the group, select one feeling (for example, elation, frustration, or boredom), and tell the two observers secretly which feeling they have selected. Then each helpee is to make verbal statements expressing, but not verbally identifying, that feeling to each of the two helpers separately. The helper is to identify the feeling. The observer will let the helper know when he or she is correct and the helpee is encouraged to continue making statements until this identification occurs.

Hearing and discriminating among affective and cognitive verbal messages without body language cues are extremely difficult and are naturally hampered by the artificiality of out-of-context role playing. Nevertheless, as you become used to role playing, you will become less artificial and more comfortable.

After you have practiced identifying cognitive and affective contents separately, try Exercise 3.19.

Exercise 3.19   In pairs (or triads), assign helpee and helper (and observer) roles. The helpee is to talk for five to ten minutes about a real or imagined concern. The helper is not to ask any questions but can make exploratory statements such as "Tell me more about that" and "I'm wondering if" to probe for more data. At the end of the agreed-on time, the helpee is to stop and the helper is to identify both the cognitive and affective messages to the helpee's satisfaction. A suggested format is "You feel _____ when _____ because _____." Allow each person the opportunity to be helpee, and then process the results of the exercise in small or large groups. The purpose of this exercise is to sharpen helpers' discrimination skills and to focus their attention on the helpee's message and the development of messages. It encourages client-centered listening in that it does not allow questions. If the role of observer is assigned in this exercise, the person playing that role should assist the helper to identify messages.

Questions usually hinder the development of the helping relationship more than they help it. We have learned to hide behind questions, and thus it is highly preferable to rephrase questions into statements until you have mastered communication skills and learned when and how to ask questions. For example, instead of asking "What did you do next?" say "Tell me what happened then."

☐ Responding Verbally and Nonverbally

Helpers must be skilled in responsive listening as a basis for responding to helpees both nonverbally and verbally. *Responsive listening* is defined as attending to the verbal and nonverbal messages and the apparent and underlying thoughts and feelings of the client. This is easier said than done and involves developing awareness of one's self as a communicator as well as refining hearing and perceiving skills.

Responsive listening implies that the helper is able to communicate his or her genuine understanding (empathy), acceptance, and concern for the helpee and, at the same time, increase the understanding of the issue by clarifying the helpee's statement. Thus, helpers must be able to communicate to the client their identification and understanding of the primary cognitive concern and the underlying feeling, as well as their own caring. It's essential that the helper be congruent in his or her own verbal and nonverbal communication, or else the helpee will be just as confused by double messages from the helper as the helper is when he or she receives double messages from the helpee.

The following is an example of responsive listening:

Helpee: I know I'm too fat. That's why nobody ever asks me out.
Helper: You really feel sad when you see everyone around you having a good time, and you're scared, wondering what will happen to you if you don't improve your appearance.

Saying "Don't worry" or "You should go on a diet" is not helpful. Those kinds of responses do not help clients increase their self-understanding. What makes the preceding excerpt an example of interpretive responsive listening is that in it the helper identifies an underlying feeling, relates it to the major cognitive concern, and adds clarity and understanding to the helpee's statement.

We'll work with more examples of responsive listening as we proceed through this chapter. Let's focus first on developing our awareness of our own nonverbal communication behaviors.

Nonverbal Responding

Exercise 3.20   The following is an important Gestalt exercise that helps us get in touch with our own comfort or discomfort with nonverbal behavior. In it, pairs sit facing each other and communicate for three to five minutes by eye contact only. No other body language or verbal language is permitted. After the time is up, the partners can continue their eye contact, but they can also communicate with their hands for three to five minutes. Then they are allowed to communicate nonverbally any way they choose. At the end of this exercise, the pairs are to process verbally, by sharing their feelings and thoughts, their intentions and reactions. Then, in small or large groups, everyone can share his or her experience. Many people find themselves very uncomfortable maintaining eye contact at first, but this form of communication becomes more comfortable and meaningful with practice and experience.

Exercise 3.21   Another significant Gestalt exercise is mirroring. In this exercise the group divides into pairs who sit facing each other. One person is the communicator; the other is the nonverbal "mirror. " The communicators talk about anything they want to for five minutes. The mirrors nonverbally mirror each gesture, movement, and expression of the communicators. (They do not verbally mirror what they think the other is saying, because they are concentrating on nonverbal behavior.) At the end of five minutes, the mirrors express their feelings, the communicators share their feelings, and they both share what they have learned from participating in this exercise. (If you do not have a partner, you can do this exercise alone by talking to yourself in front of a mirror.)

Because we rarely see ourselves communicating, we are generally unaware of our nonverbal behaviors. However, the mirror exercise allows us to observe and discuss whether our nonverbal behaviors are facilitative or nonfacilitative. For example, one student discovered that by sitting back and rocking in his chair, he was distancing himself from others and not being facilitative.

Exercise 3.22   In triads, the helpee communicates a real or imagined stress or crisis. The helper attempts to respond to these messages, and the helpee and the observer give direct feedback to the helper about his or her nonverbal behavior. Observers can refer to the section titled "Nonverbal Behaviors" in Appendix A as a guide to evaluating and developing helpers' awareness of nonverbal behaviors.

Desirable nonverbal behaviors for effective helping relationships include occasional nodding, smiling, touching, and hand gesturing; good eye contact; facial animation; leaning toward the helpee (sitting near, without a desk barrier); a moderate rate of speech; and a firm, supportive tone of voice. Nonverbal behaviors communicate warmth, understanding, attentiveness, and efficacy, apart from and in congruence with verbal behavior.

Verbal Responding

When we respond verbally, we attempt (1) to communicate to helpees that we are truly hearing and understanding them and their perspective; (2) to communicate our ability to help, our warmth, acceptance, and caring; and (3) to increase the client's self-understanding and self-exploration as well as his or her understanding of others by focusing on major themes, clarifying inconsistencies, reflecting back the underlying feelings, and synthesizing the major apparent and underlying concerns and feelings.

We have already begun to focus on verbal responding by identifying the major affective and cognitive contents of helpee statements. We are also developing the ability to generate additive (facilitative) responses, which help the client to understand his or her thoughts and feelings and add some understanding of what the client is trying to communicate. This form of responsive listening is particularly helpful during the relationship stage of helping, because it facilitates the objectives mentioned above. Responses that are simply reflections or paraphrases of the client's verbalized and nonverbal thoughts and behaviors are considered interchangeable, and responses that do not pertain at all to the client's message are subtractive, or nonfacilitative.

Following these guidelines will help you reach the goals of verbal responding just listed.

1. Listen to the helpee's basic message.
2. Respond to the most important part of the helpee's statement that coincides with the basic apparent or underlying verbal and nonverbal messages.
3. Reflect the helpee's feelings at a greater level of intensity than originally expressed by him or her.
4. Reflect both implicit and explicit feelings of the helpee and help him or her differentiate between thoughts and feelings.
5. Respond to the helpee's behaviors.
6. When the helpee changes topics, respond to the primary cognitive or affective theme of the topics.
7. Always allow the helpee to modify or reject your perceptions.
8. Use your own feelings as the basis for checking out, confronting, **leading,** and so on.
9. If you are unable to rephrase your questions as statements, ask only open-ended questions that clarify issues for the helpee or that elicit feelings.
10. If the helpee doesn't pause to give you a chance to respond, and you feel lost or confused, break in with a statement such as "I feel confused . . ." in order to focus on major themes. But don't feel you have to respond to every single helpee statement.

The next exercise will give you an opportunity to identify responses that are interchangeable, subtractive, or additive in terms of communicating your goals to the helpee. At the same time that you learn to differentiate between facilitative and nonfacilitative responses, you will learn, through **modeling,** *what facilitative responses are.*

Exercise 3.23   Rate each of the following statements as either – (subtractive: does not increase the client's understanding), = (interchangeable: neither increases nor decreases the client's understanding), or + (additive: increases the client's understanding). Each rating may appear more than once in each numbered set.

1. **Client:** I need to find a job quick. I've got a family to support. The economy is terrible.

Helper:

___ a. Why did you leave your last job?

___ b. You really are scared about how you're going to make ends meet.

___ c. You want to find a job right away so you can take care of your family.

In this situation, the first response would be rated − because it asks a "why" question and puts the client on the defensive, not adding to his or her understanding of the problem. Response b is rated + because it identifies the underlying feeling and communicates empathy. Response c is rated = because it neither increases nor decreases understanding but merely paraphrases the message. Now finish rating the statements and then check the answers at the end of the chapter.

2. **Client:** Put the flowers over there and the newspaper there, and bring me my slippers and robe, please.

   Helper:

   ___ a. You really want me to be able to give you more attention.

   ___ b. You want me to put down your flowers and newspaper so I can bring you your slippers and robe.

   ___ c. Here you go again. I have other patients to care for, too!

3. **Client:** I'm not going to be able to come back to school next year.

   Helper:

   ___ a. Why not?

   ___ b. People who graduate have a better chance of getting jobs.

   ___ c. You're not coming back for your junior year.

4. **Client:** Yeah, well, you gotta be tough if you're gonna make it on the outside. Nobody gives an ex-con a chance anyway, so you gotta take what you can get and see that you get it.

   Helper:

   ___ a. You really feel that no one's going to give you a break.

   ___ b. You're wondering if you're tough and smart enough to make it. It's scary.

   ___ c. Ex-cons don't have a very good track record, you know.

5. **Client:** When I'm at home, my mom lets me eat whatever I want.

   Helper:

   ___ a. Wow! Your mom sure spoils you!

   ___ b. You don't want to have to eat what you don't like.

   ___ c. You're unhappy that you can't always do what you want to here.

6. **Client:** I'm so angry at my mother that I'd like to kill her! She never says a nice thing to or about me. I wish I never had to see her again.

   Helper:

   ___ a. It makes you really angry that you seem to need her approval, doesn't it?

   ___ b. It's wrong to even think about your mother like that.

   ___ c. You're really angry with your mother. Tell me more.

7. **Client:** This is an awful place to work. No one is ever where they should be.

Helper:

___ a. You don't like working here.

___ b. All offices are like that in this kind of business.

___ c. It seems to you that nobody cares about you here and that nobody wants to help you.

8. **Client:** It's unfair that you can't find more money for me. How am I supposed to manage? I've got a wife and four kids.

   Helper:

___ a. You're concerned about making ends meet.

___ b. You're getting as much as you're entitled to under the law.

___ c. Why do you think you should get more than anyone else?

9. **Client:** People today care more about money than they do about one another.

   Helper:

___ a. You feel lonely and scared that people don't seem to care about you.

___ b. It makes you angry that people are so materialistic.

___ c. Yes, that's the world we live in today.

10. **Client:** It's a lousy course. I've had all that stuff before and it's a waste of my time and money. Most of the courses around here are pretty bad.

    Helper:

___ a. You're not sure you should be in that course.

___ b. It's a required course for this program.

___ c. You seem to have ambivalent feelings about being in this program. Can we talk about that?

11. **Client:** I'd love to go back to work, but my husband feels I should be home when the kids get back from school.

    Helper:

___ a. I can see why he feels that way. It's much better for the children when their mom is home.

___ b. You're not sure whether to work or stay home.

___ c. Sounds like you feel some anger toward your husband because he imposes his expectations on you.

12. **Client:** I've really had a rough year.

    Helper:

___ a. You've had a tough time this year.

___ b. Everyone has a bad year at some time or another.

___ c. You seem to be pretty uptight about how you've handled things this year.

13. **Client:** Well, he's got a hell of a nerve telling me what to do in that tone of voice. Who does he think he is?

    Helper:

___ a. Bosses are known to do that. Don't take it to heart. I'm sure his bark is worse than his bite.

___ b. You sure get angry when someone pushes you around!

—   c. You're really angry that he doesn't treat you with respect and accept you as a person who has feelings.

14. **Client:** Look, I'm only here because Mr. Smith sent me. I've got nothing to talk about.
    Helper:
—   a. Mr. Smith wanted you to come see me.
—   b. You don't want to be here and you're angry that you got yourself into this.
—   c. He must have had some reason for sending you here.

15. **Client:** Every night my wife complains about everything that's happened during the day. It's getting so I don't want to go home anymore. I'd much rather stay in town and drink with the boys.
    Helper:
—   a. All wives are like that. After all, what else have they got to do?
—   b. You find it so intolerable at home that it's easier for you to stay away. Sounds as if you're pretty angry at your wife.
—   c. Your wife really rides roughshod on you when you get home every night.

Note that the nonfacilitative responses in Exercise 3.23 neither increase the client's understanding of the problem nor focus on the underlying feelings. Rather, they tend to moralize or preach and avoid affective parts of client statements. The interchangeable responses do not close the door for further development, but they do not add to what has already been stated. Helpers who continually make interchangeable responses need to examine carefully their own avoidance behavior. Are they afraid to risk testing their understanding? Facilitative responses communicate the helper's listening, understanding, and caring; they help focus on the implicit and explicit affective and cognitive content; and they may encourage further exploration. A helper's response that incorrectly identifies the client's underlying feeling would still be considered facilitative. As long as the client has the opportunity and encouragement to say something like "No, not that, but . . ." further exploration is facilitated. So it is not so much a question of a right or wrong response as of how facilitative the response is in terms of empathy, honesty, and open-endedness.

At this point, you may be thinking to yourself, "But isn't this kind of verbal responding putting ideas and thoughts into the client's head?" The answer to that question is that genuine, client-centered, empathic responses are your best insurance against putting words, ideas, or your values or needs into your client's head, because making those kinds of responses means that you are hearing and understanding client messages and, at the same time, continually allowing for feedback and reactions to your responses. By your manner and responses you can communicate to clients your respect and confidence in their ability to think, feel, and act for themselves. You neither want nor need to do that for them. If the client doesn't like or agree with what you say, that's fine; it doesn't mean that you are no good or that the client is resisting you. It does mean, however, that you can both continue to explore what is going on until you can agree on what the problem is all about. It is a good idea to remember that timeliness and individual variables will influence your choice of responses. Some clients may become defensive if they receive an additive response too early in the relationship.

The next exercise asks you to write a facilitative response to each client statement. Check your responses against mine at the end of the chapter to see if you're in the ballpark. There is no one right response to any statement. Discuss your responses with others in your group.

Exercise 3.24   For each helpee statement, write the best verbal response you can think of to meet the communication goals discussed in this chapter. Remember that these statements are out of context, and therefore you can respond only to verbal cues.

1. **Client:** I had a great time last night. I didn't think about Dave one minute!
   Helper: *(possible answer)* You're relieved that you were able to have fun and not think about Dave.

2. **Client:** John was snorting coke at the party and he wanted Tom and me to do it, too. But I was scared we'd get into trouble.
   Helper:

3. **Client:** We'd like to get married, but we know we have many problems. What do you think we should do?
   Helper:

4. **Client:** You know, I wrote that financial report, but because I'm only an assistant I can't even get credit for it.
   Helper:

5. **Client:** We don't need his folks' help. We can do it ourselves, and I wish Jack would realize that his mother's always butting into our affairs.
   Helper:

6. **Client:** I refuse to let my kid be sent to that school with all those kinds of kids. She's gonna stay right here where she belongs.
   Helper:

7. **Client:** I'm not so sure that I can handle this job. It may be too much for me.
   Helper:

8. Client: Listen, mister, you better believe that once I get outta here, there's no way you're gonna get me back. I'll die first and I'll take some of your kind with me, you wait and see.
Helper:

9. Client: The boys won't let me play ball with them. They're always teasing me and calling me names. I hate them!
Helper:

10. Client: I think people are two-faced.
Helper:

11. Client: You're always late. I've got more important things to do than sit around your office waiting for you, you know.
Helper:

12. Client: I can't take that test. I have a splitting headache. Will you talk to Ms. Smith for me?
Helper:

13. Client: I didn't do nothin'. You're always picking on me.
Helper:

14. Client: I made it through school on my own. Why should I pay his tuition? Let him work like I did.
Helper:

15. Client: I'm always telling Jim not to argue with his father. His father has a terrible temper.
Helper:

16. Client: I'm so mad at my boss! I'd like to wring his neck. I do all the work around here and he doesn't even recognize that.
Helper:

17. Client: How can he expect me to work, take care of the house, and raise the children? He better find work pretty soon so we can afford some help.
Helper:

18. Client: I managed just fine until the accident. I'm blind now, and I just have to face the fact that I can't do what I used to do.
Helper:

19. **Client:** Other people have no idea how expensive it is to care for a handicapped child. We have to keep borrowing from my family.
    Helper:

20. **Client:** I now have a job, Johnny is in day care, and I just can't believe how wonderful everything is. For the first time in 44 years, I feel like a whole person!
    Helper:

Now that you have had an opportunity to respond in writing to client statements, it is time to try out verbal responses. The next few small-group exercises allow you to verbalize responses that demonstrate responsive listening. A useful adjunct to these exercises is to tape-record sessions with actual helpees or with a friend or member of your family and then analyze your tape. Remember that this is a learning process, that putting concepts into practice is very difficult, and that it will take continual practice and time to achieve effective responsive listening skills.

Exercise 3.25   In this exercise, you should divide into triads. (It is a good idea for these triads to remain the same for the duration of the course so that, as trust develops within the group, you will feel freer to voice real concerns rather than role play.) Each triad meets for at least one hour per week, in or out of class. Each member of a triad should have the opportunity to be helper, helpee, and observer for at least 15 minutes at each meeting. The helpee may present a real personal issue or concern (it does not have to be a crisis or even a negative issue) or role play a concern. The helper demonstrates effective responsive listening skills. Problem solving and solution giving are to be avoided. The purpose of these interactions is for helpees to express their concerns and for helpers to make helpees feel understood and to facilitate further exploration. The observers are free to break in if they feel the helper is getting off track or getting into problem solving. At the end of about 15 minutes, the triad processes their practice counseling session. It is as important for observers to give honest feedback to helpers as it is for helpees to share their reactions and feelings. You may find that you best learn the significance of effective and noneffective listening when you play the role of helpee. Helper and helpee should discuss whether they felt they were on the same wavelength at the same time.

With regard to observer ratings, Kagan (1980) has found what is called *interpersonal process recall* to be helpful in processing counselor interactions. In this process, the observer asks the following questions of the helper when processing the triad sessions:

1. What do you think the helpee was trying to say?
2. What do you think the helpee was feeling at this point?
3. Can you pick up any clues from the nonverbal behavior?
4. What was running through your mind when the helpee said that?

5. Can you recall some of the feelings you were having then?
6. Was there anything that prevented you from sharing some of your feelings and concerns about the person?
7. If you had another chance, would you like to say something different?
8. What kind of risk would there have been if you had said what you really wanted to say?
9. What kind of person do you want the client to see you as being?
10. What do you think the client's perceptions of you are?

The observer can also use a rating scale to aid in providing feedback to the helper, or he or she can make audio or video tapes of the session and then analyze the playback. The rating scale focuses on the verbal and nonverbal behaviors desired in helping situations and enables assessment of the level of communication the helper uses. In general, overparticipation occurs when helpers feel they must say something at every pause and put much of their own energy into filling gaps. Adequate participation occurs when the energy levels of the helper and helpee are about equal and the helper feels comfortable with occasional silences and pauses. Finally, underparticipation occurs when helpers are so insecure about their verbal responses that they allow helpees to go on and on and never intervene. In the last case, the helpee has a higher level of energy invested in the communication than the helper. Once you become aware of your own patterns and style of communication you will be able to modify your responses in the best interests of the helpee.

Exercise 3.26   This is a variation on Exercise 3.25; however, in this case, small groups of six to ten people sit in a circle. One person volunteers to be the helpee and begins to communicate a personal concern. Each time the helpee makes a statement, a different person in the circle makes a facilitative response, going around the circle. For example, the helpee makes an opening statement and person A responds; then the helpee replies to person A and person B responds to that, and so forth around the circle. This exercise is processed in a manner similar to Exercise 3.25. It has the advantage of involving more people in the processing and producing various levels of responses to the same person. Thus, the helpee is able to provide valuable feedback: whose responses were most helpful, why, and whose were not helpful. Actually, the most important benefit of this exercise is for the helpee, in that he or she is able to feel the effects of "connected" and "unconnected" responses.

Exercise 3.27   Read the following statements. Try to imagine that the person is speaking directly to you, the helper. Then write a response that (1) includes your understanding of both feelings and content and (2) will encourage the helpee to continue talking. Statement number 1 serves as an example.

1. **30-year-old woman:** I'm really upset. My parents keep needling us about having a baby. Joe and I aren't sure that we want to have children. They can really disrupt your life. Most of our friends who did have kids really got bummed out. Anyway, my parents keep wanting us to come

home for the holidays, but every time we do, it gets really tense. I wish they'd leave us alone.

Helper: You really are concerned about how you and Joe can lead your own lives and make your own decisions without hurting your parents.

2. **Female college student:** I don't understand why grades are so important. College should be a place to have fun, but my folks are always on my back about my grades. They say if I ever want to go to graduate school, "da da da dum." They really get annoyed if I'm having a good time.
   Helper:

3. **14-year-old boy:** I'm really worried about my older sister [age 16]. She thinks I don't know what's going on, but I know she's into drugs and that she's screwing with her boyfriend. And the folks act like nothing's going on. She's moody and rotten to everyone at home.
   Helper:

4. **Nursing student:** It's impossible to spend any time talking to patients. The head nurse is always bickering and telling us to move on, the docs are always changing their orders, and we really get treated worse than the orderlies. Sometimes I wonder why I ever thought I wanted to become a nurse.
   Helper:

5. **Secretary:** I was supposed to meet my brother for dinner tonight. But Ms. Green asked me to work late tonight to get this report finished, and she's been so good to me I don't see how I can let her down.
   Helper:

6. **39-year-old business executive:** There's never enough time to get everything done. Business is slow and everyone is on edge around here. There are rumors that maybe payroll can't be met. And I just found out my kid needs braces and my wife wants the kids to go away to camp next summer. I don't know how I'll pay my heating bills this winter.
   Helper:

7. **22-year-old college graduate:** I have been on umpteen interviews. I know exactly what kind of job I want and I can't seem to get it. Someone else will have already gotten there or there just isn't anything open now. I don't know where else to look or what to do, and I'm going stir crazy sitting around the house waiting for people to call me back.
   Helper:

8. **63-year-old woman:** My children never come to see me. They're always too busy. When I call to talk to them, they always seem rushed. I don't un-

derstand it. I was always good to my mother. I spent the best years of my life being a mother, and this is what I get for it.
Helper:

Advanced Verbal Response Skills

After you have learned to recognize facilitative and nonfacilitative responses, you can begin to develop patterns of verbal responding congruent with the types of issues involved and the stage of the helping relationship. Ten of the most commonly used kinds of verbal response are making the minimal verbal response, paraphrasing, probing, reflecting, **clarifying,** checking out, **interpreting, confronting,** informing, and summarizing.

> *Making the minimal verbal response:* Minimal responses are the verbal counterpart of occasional head nodding. These are verbal cues such as "mm-mm," "yes," "I see," "uh-huh," which indicate that the helper is listening and following what the client is saying.
>
> *Paraphrasing:* A paraphrase is a verbal statement that is interchangeable with the client's statement, although the words may be synonyms of words the client has used. For example:

Client: I had a lousy day today.
Helper: Things didn't go well for you today.

> *Probing:* Probing is an open-ended attempt to obtain more information about something and is most effective when done using statements such as "Tell me more," "Let's talk about that," "I'm wondering about . . ." rather than "how," "what," "when," "where," or "who" questions.
>
> *Reflecting:* Reflecting refers to communicating to the helpee our understanding of his or her concerns and perspectives. We can reflect stated or implied feelings, what we have observed nonverbally, what we feel has been omitted or emphasized, and specific content. Examples of reflecting are "You're feeling uncomfortable about seeing him," "You really resent being treated like a child," and "Sounds as if you're really angry at your mother."
>
> *Clarifying:* Clarifying is an attempt to focus on or understand the basic nature of a helpee statement. Examples are "I'm having trouble understanding what you are saying. Is it that . . . ?" "I'm confused about . . . Could you go over that again, please?" and "Sounds to me like you're saying . . ."
>
> *Checking out:* Checking out occurs when the helper is genuinely confused about his or her perceptions of the helpee's verbal or nonverbal behavior or when the helper has a hunch that bears trying out. Examples are "I feel that you're upset with me. Can we talk about that?" "Does it seem as if . . . ?" and "I have a hunch that this feeling is familiar to you." The helper asks the helpee to confirm or correct the helper's perception or understanding, in

contrast to a clarifying request, which elicits a deeper, clearer understanding.

Interpreting: Interpreting occurs when the helper adds something to the client's statement or tries to help the client understand his or her underlying feelings, their relation to the verbal message, and the relation of both to the current situation. For example:

Client: I just can't bring myself to write that report. I always put it off and it's hanging me up right now.

Helper: You seem to resent having to do something you don't want to do.

If the interpretation is useful, it will add to the client's understanding, and you will receive a reaction reflecting "Yes, that's it." If it's not useful, the client may say "No, not that but . . ."

Confronting: Confronting involves providing the helpee with honest feedback about what is really going on. The confrontation may focus on genuineness, reflected in statements such as "I feel you really don't want to talk about this," "It seems to me you're playing games in here," and "I'm wondering why you feel you always have to take the blame. What do you get out of that?" Or the confrontation may focus on discrepancy, reflected in statements such as "You say you're angry, yet you're smiling," and "On the one hand you seem to be hurt by not getting that job, but on the other hand you seem sort of relieved, too." An effective way of using confrontation is to send "I" messages, to "own" your responsibility for the confrontation by openly sharing your own genuine responses to the helpee or by focusing on the helpee's avoidance or resistance.

Informing: Informing occurs when you share objective and factual information such as what you know about a particular college in terms of student enrollment, types of programs, and so on. It's important for the helper to separate informing from *advising,* which is subjective and verges on telling the helpee what to do. Advice is all right as long as it is tentative, with no strings attached, and as long as it's clearly advice, not a demand.

Summarizing: By summarizing, the helper synthesizes what has been communicated during a helping session and highlights the major affective and cognitive themes. Thus, a summary is a type of clarification. This response is important at the end of a session or during the first part of a subsequent session. Summarizing is beneficial when both the helper and the helpee participate and agree with the summary. It also provides an opportunity for the helper to encourage the helpee to share his or her feelings about the helper and the session.

The following are general guidelines for giving the ten major kinds of verbal responses:

1. Phrase your response in the same vocabulary that the helpee uses.
2. Speak slowly enough so that the helpee understands each word.
3. Use concise rather than rambling statements.

Table 3.2 Role Behaviors and Verbal Responses Appropriate to Stages of Counseling

	Stage	Role Behavior	Verbal Response
Responsive Listening	Relationship	Attending Clarifying Informing/describing Probing/inquiring Supporting/reassuring	Minimal verbal response Paraphrasing Probing Reflecting Clarifying Checking out Informing
	Strategies	Attending Informing/describing Probing/inquiring Supporting/reassuring Motivating/prescribing Evaluating/analyzing Problem solving	Probing Checking out Interpreting Confronting Informing Summarizing

4. Pursue the topic introduced by the client.
5. Talk directly to the client, not about him or her.
6. Send "I" statements to "own" your feelings, and allow the client to reject, accept, or modify your messages.
7. Encourage the client to talk about his or her feelings.
8. Time your responses to facilitate, not block, communication.

The ten major kinds of verbal response may be applicable throughout the helping relationship and/or during particular stages. For example, interpreting, confronting, and informing may be more appropriate during the strategies (working) stage, whereas minimal verbal response and paraphrasing may facilitate beginning relationship sessions or interviews.

Doyle (1982) discusses eight "communication role behaviors"—that is, communication behaviors linked to a specific role—as contexts for communication skills: (1) attending, (2) clarifying, (3) informing/describing, (4) probing/inquiring, (5) supporting/reassuring, (6) motivating/prescribing, (7) evaluating/analyzing, and (8) problem solving. We can relate these role behaviors and the ten major kinds of verbal response to the two stages of counseling, as seen in Table 3.2.

Exercise 3.28 The purpose of this exercise is to see if you can recognize and identify the major types of verbal responses just discussed. This exercise will help you become aware of your own verbal responses and perhaps encourage you to expand your repertoire. Read the following client and counselor statements and then identify the counselor's response in each case as making the minimal verbal response, paraphrasing, probing, reflecting, clarifying, checking out, interpreting, confronting, informing, or summarizing.

1. **Client:** That's why I'm here. Jim said you were a good one to talk to.
   Helper: Let's see now. You want me to help you decide whether or not you should accept the transfer. Is that right?
2. **Client:** Do you think they have a good benefits package there?
   Helper: The National Conference Board reports that that particular company ranks in the top third for employee benefits.
3. **Client:** Eddie made me get kicked out of class today.
   Helper: Tell me more about it.
4. **Client:** I have to get the house cleaned before we can have company over.
   Helper: I see.
5. **Client:** In my family, my dad and brother don't do any of the work around the house.
   Helper: The men in your family don't do any housework.
6. **Client:** I can't decide what to do. Nothing seems right.
   Helper: You're feeling pretty frustrated and you want me to tell you what to do.
7. **Client:** I don't want to talk about it.
   Helper: You always seem to back away when things get personal. It seems to me that it's much easier for you to talk about the situation than feel it.
8. **Client:** Anyway, I'm unable to do it because it's too expensive and besides they won't help me anyway.
   Helper: Let me get this straight. You feel the tests will cost too much and the results won't be worth the cost. Is that it?
9. **Client:** Nobody in this world cares about anyone else.
   Helper: It's scary to feel that nobody at all cares about you.
10. **Client:** I guess that about covers it.
    Helper: Let's see if we can review what we've talked about today. . . . Does this seem right to you?

☐ Summary

In this chapter, we have discussed the cognitive and affective components of verbal and nonverbal messages, identified and elaborated the elements of responsive listening, and outlined the ten major kinds of verbal response. A progressive series of individual and group exercises was presented, first to develop awareness of your own style of nonverbal and verbal communication behaviors, then to test your understanding of the concept of responsive listening—the differences among subtractive, interchangeable, and additive responses—and finally to allow you to practice and develop more effective communication skills.

We are focusing our exercises and work on responsive listening as the core communication skill. It is also the most difficult to learn and master, in that it is the least commonly used form of communication in our society. Not only is responsive listening essential to establishing rapport and attending to the helpee's verbal and nonverbal messages, but it is also helpful in identifying and clarifying the helpee's underlying concerns. We have also described ten commonly used forms of verbal responses that can be used in connection with overall responsive listening in the two stages of the helping relationship.

Guidelines are given in this chapter for developing communication skills. The effectiveness of the exercises will depend largely on the supervisory and modeling capabilities of your supervising trainer, but an instructor cannot spend a great deal of time supervising any one individual. Members of a group, however, can learn to be effective observers and provide one another with beneficial instruction in the form of honest feedback. Communication skills are the fundamental basis of the helping relationship and can be learned and practiced in many different formats, so you will have further opportunities to develop them in the chapters to come.

☐ Exercise Answers

Exercise 3.1   Possible answers might be (1) openness, (2) defensiveness, (3) evaluation, (4) rejection, (5) rejection, (6) helplessness, (7) embarrassment, (8) self-control, (9) boredom, (10) acceptance.

Exercise 3.10
   1. a. I don't have enough money.
       b. I have bills to pay.
       c. Should I take another job?
       d. Should I go back to school?
   2. a. Is school worth the money it costs?
       b. I'm not learning anything here.
       c. People who go to school don't get jobs later.
       d. People get lost in such a big, impersonal place.
   3. a. I don't know what to say to Mr. Jones when he calls later.
       b. Somebody ought to fix the parking situation.
       c. I need to be able to park in order to get here on time.
       d. I lost a memorandum.
       e. There aren't enough parking places here.
   4. a. My apartment is not satisfactory because there is no heat.
       b. My kid is sick.
       c. I don't know what to do or where to start.

d. I need money to pay the bills.
e. I need a job.
f. I am alone here.
g. I just moved here.
h. I have not received all my furniture.
i. I have a disagreeable landlord.
5. a. We had a great time.
   b. The Meyers were down there.
   c. through f. could be interchangeable: swimming (c), bicycling (d), fishing (e), sailing (f).

Exercise 3.12   1. a, 2. c, 3. a, 4. a or c, 5. c, 6. all, 7. c, 8. c

Exercise 3.16

1. Fear of bullies
2. Fear of illness or pleasure at attention from doctor
3. Sympathy for Lenny or anger at Lenny for never getting home
4. Anxiety about coming exams or anticipation of vacation after exams
5. Feeling pressured and rushed or bitter and angry about not being able to take a coffee break
6. Anger at being overworked by Professor Greene or distressed at being too busy to type report
7. Frustration and anger or loneliness
8. Fear of being hurt or anger at others' selfishness
9. Exasperation and frustration about delay or anger at dependence on Jim
10. Anxiety about new social worker or excitement about meeting her
11. Fear
12. Anticipation
13. Anger at Ms. Jones or concern for Ms. Jones
14. Desperation
15. Fear that something is wrong with child
16. Anger at young people for having more sexual freedom, or jealousy
17. Relief at stepping down or bitterness about pressures
18. Frustration at inability to learn
19. Concern
20. Confusion, fear
21. Anger at being ignored
22. Discouragement about not getting anywhere or anger at helper for not being better helper
23. Anger at rejection, or loneliness
24. Fear of rejection, or loneliness
25. Anger or concern

Exercise 3.23

2. a. +	3. a. −	4. a. =
b. =	b. −	b. +
c. −	c. =	c. −
5. a. −	6. a. +	7. a. =
b. =	b. −	b. −
c. +	c. =	c. +
8. a. +	9. a. +	10. a. =
b. −	b. =	b. −
c. −	c. −	c. +
11. a. −	12. a. =	13. a. −
b. =	b. −	b. =
c. +	c. +	c. +
14. a. =	15. a. −	
b. +	b. +	
c. −	c. =	

Exercise 3.24 Possible helper responses might be the following:

2. "Sounds to me as though you were annoyed.at them for tempting you."
3. "You're confused about whether or not your marriage will work if you haven't worked through your problems."
4. "It makes you feel angry and inadequate when you're not given credit for your work."
5. "You're pretty angry at Jack for not being able to wean himself away from his folks."
6. "You're afraid something will happen to her and you care so much."
7. "You're afraid that you may be inadequate and may fail on this job."
8. "I hear how determined you are to make a go of it. That's terrific."
9. "It makes you feel sad and lonely when they won't let you play with them."
10. "You've been hurt often and you're afraid to trust people."
11. "You feel like I don't care enough about you if I keep you waiting."
12. "You're so afraid you won't do well on that test."
13. "It seems to you that I don't like you, that I'm unfair to you."
14. "You're angry that he wants your help when you were strong enough to do it on your own."
15. "You're afraid of anger."
16. "You don't feel appreciated."
17. "You're really scared and overwhelmed with responsibilities."
18. "You're very independent and you seem determined and proud."
19. "You're very angry that you have to be dependent on your family."
20. "Things are finally going well for you and you don't want anything to happen."

Exercise 3.28

1. checking out
2. informing
3. probing
4. making the minimal verbal response
5. paraphrasing
6. reflecting
7. confronting
8. clarifying
9. interpreting
10. summarizing

☐ References and Further Reading

Benjamin, A. (1981). *The helping interview* (3rd ed.). Boston: Houghton Mifflin.

Brammer, L., Shostrum, E., & Abrego, P. J. (1989). *Therapeutic psychology: Fundamentals of counseling and psychotherapy* (5th ed.). Englewood Cliffs, NJ: Prentice-Hall.

Cormier, L. S., & Hackney, H. (1987). *The professional counselor: A process guide to helping.* Englewood Cliffs, NJ: Prentice-Hall.

Cormier, W. H., & Cormier, L. S. (1985). *Interviewing strategies for helpers: A guide to assessment, treatment, and evaluation* (2nd ed.). Pacific Grove, CA: Brooks/Cole.

Doyle, R. E. (1982). The counselor's role communication skills, or the roles counselors play: A conceptual model. *Counselor Education and Supervision, 22,* 123–132.

Evans, D., Hearn, M., Uhlemann, M., & Ivey, A. (1984). *Essential interviewing* (2nd ed.). Pacific Grove, CA: Brooks/Cole.

Gordon, T. (1970). *Parent effectiveness training.* New York: Wyden.

Ivey, A. (1972). *Microcounseling: Interviewing skills manual.* Springfield, IL: Charles C Thomas.

Kagan, N. (1980). Affect simulation in interpersonal process recall. *Journal of Counseling Psychology, 16,* 309–313.

Kagan, N. I., & Kagan, H. (in press). *IPR—A research/theory/training model.* In P. W. Dowrick (Ed.), *A practical guide to video applications in psychology.* New York: Grune & Stratton.

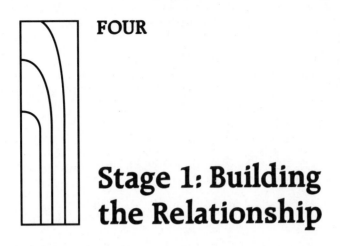

FOUR

Stage 1: Building the Relationship

In Chapter 1, we posited the development of the helping relationship as crucial to the effectiveness of any helping strategy. We then turned to a discussion of successful helper characteristics and communication skills. We move on now to examine the conditions and steps necessary for creating the type of empathic climate in which the helpee can begin to explore his or her world and gain self-awareness.

☐ Conditions Affecting the Relationship Stage

Before discussing the five steps of the relationship stage, let's look at the inherent conditions affecting this stage. These conditions may appear obvious to you, but unless they are taken into account consciously, they may hinder the helping relationship.

Initial Contact

The term *interview* is often used to describe the first one or two helping sessions, because these sessions are usually for information gathering. For some people, the word has threateningly formal connotations: we can all think of job and school interviews we've endured that have been more like inquisitions. Whether the interview is initiated by the helper, the helpee, or a third party, the helpee often feels anxiety about being accepted and fears saying the "wrong" things.

Nonetheless, we'll use the term *interview* for the initial meetings when the participants first meet and collect information. If the helping relationship proceeds, subsequent meetings can be called *sessions*. The first interview is really a testing period for both parties: helpers ask themselves if they will be able to work effectively with the helpees, and helpees ask themselves if they will be able to trust and respect their helpers and disclose their real concerns to them. An important aspect of the first interview is discussion of both parties' expectations of each other and of the helping process.

The tone of an interview is set at the first moment of initial contact, whether that be when the appointment is made or when the interview begins. First the helper and the helpee must establish a mutually convenient meeting time. Right away, the helper should be frank. If unable to devote adequate time to the helpee at that particular moment, rather than trying to rush through a few frantic exchanges on the phone or in a corridor, the helper is better off saying, "I see you really need to talk to me about that. I'm tied up now. Could you come (call) back at three o'clock?"

The helper must quickly determine the realities and priorities of any given situation: Is this a crisis? Can I rearrange my schedule? Can this person wait until later to see me? Sometimes the only way to determine the nature of a situation is to ask helpees whether or not they can wait. In any case, at this critical moment of initial contact the helper can communicate genuine concern and willingness to be available to the helpee. The following are some points to remember when arranging an appointment.

1. Schedule a specific time for an appointment; avoid saying "later," "next week," or "soon," and suggest specific alternatives.
2. Tell the helpee that you want to be available to him or her when you both can give your full time and attention to the helpee's concerns.
3. Communicate support (reinforcement) to the helpee for initiating contact (for example, "I'm so glad you've come to see me about this"). People often feel a bit foolish and unsure about asking for help, and they need reassurance.

If you work in a setting where appointments are made for you, you must assume responsibility for seeing that whoever makes appointments communicates the same kind of concern and helpfulness to the client. There is nothing more deflating to a person seeking help than to be rudely put off by an appointment maker who delays appointments, refers to how busy you, the helper, are, or is too inquisitive and asks too many personal questions.

If at all possible, arrange to have a corner where people waiting to see you may be unobserved by others in your setting. Many people feel uncomfortable if a peer knows they are talking to a supervisor, a counselor, or some other helping person.

If you are initiating the first interview, tell the helpee why you are arranging this meeting, preferably at the appointment-making time or at the beginning of the first contact. It is necessary that the helpee understand why you've arranged the contact so that he or she can learn to trust you and in order to offset his or her resistances and defenses.

Duration

A helping relationship can last one, a few, or many sessions. The number depends on the following:

1. The nature of the relationship (whether it is formal or informal, voluntary or involuntary, and so on)
2. The nature of the problem (whether it is short-term or long-term; how easily it is defined, clarified, and accepted; whether it is a crisis or preventive or developmental)
3. The setting in which the relationship occurs (whether it is a counseling center, a human services institution, a business, and so forth)

In many cases, it is unlikely that the nature and duration of the helping relationship can be predetermined until the nature of the problem is clarified. Thus, most initial sessions begin in a similar fashion, with the dual purpose of having the involved parties establish rapport and identify the problem.

Applications and Forms

If applications or forms are necessary to the helping process, it is usually better to ask the helpee to complete them before you begin the interview so that the meeting time can be used to establish a positive relationship. However, if your time allows, aiding the helpee to fill out the forms can be useful in establishing the helping relationship. Just remember not to become so involved in the content of the forms that you miss what the helpee is really communicating.

Facilities

To establish trust and support, it is necessary to provide a meeting place where confidentiality can be ensured. Conferences in open offices with thin partitions or in the corner of an occupied room can be difficult situations in which to uncover a client's real concerns. If a private room is not available, you can try to find an out-of-the-way section in a stairwell or corridor, or go out of the building.

If you do have a private office, try to arrange the furniture so that you can sit facing the helpee without any barriers between you. For example, you can sit on one side of your desk close to the helpee, rather than across from him or her. Some people prefer to work away from their desks and arrange chairs facing each other in another part of the room.

Timing

Except in unusual circumstances, try to keep appointments as scheduled. It is terribly frustrating when someone comes to see you and finds you are not available.

You can let the helpee know just how much time you have at the outset, if this is not already understood. During all interviews and sessions, it is important that you demonstrate your involvement with the helpee by refusing to accept telephone calls and by not allowing interruptions of any kind. A common complaint voiced by people seeking help is that they had just begun to get down to their real concerns when the mood was interrupted.

Another aspect of timing is a consideration of the time of day or evening when both helper and helpee function best. Some people are aware of being more alert during certain periods of the day than during others.

Record Keeping

If you intend to take written notes or use a video or audio tape recorder during the interview, you can begin by explaining your rationale for this procedure and by clarifying who will and will not have access to the recorded data. Do not assume that a helpee who doesn't say anything about the record keeping doesn't have any questions or feelings about it. You should bring it up and frankly discuss what it is all about. Both helper and helpee usually become oblivious to recording devices after the first few moments. Very often, helpees ask to review video and audio tapes, which can prove to be an excellent strategy for developing both the helpee's and the helper's self-awareness. Taking notes during sessions or interviews can be distracting and can impair helper and helpee attentiveness; if you must do so, try to be brief and unobtrusive.

Your agency may have an informed consent form for the helpee to sign. If not, you may want to consider developing your own. See Appendix C for a model of this kind of form.

Other People

Sometimes a helpee will bring someone else such as a friend, relative, or interpreter along to the interview. According to the nature of the situation and setting, you will have to decide whether the presence of another person will help or hinder the helping relationship, and act accordingly. When an interpreter is necessary, remember that the communication process between you and the helpee is filtered through another layer and that the relationship between the interpreter and helpee and between you and the interpreter is never neutral. When the other person is a friend or relative, the opportunity for observing and hearing differing or new perspectives about the helpee and her problems can be considered.

Some of the preceding points fall under the heading of obvious common courtesy. Yet we can all think of situations when we, as helpees, have been stymied by a helper's failure to consider one or more of the conditions affecting the relationship stage.

Think back to interviews you have participated in. What conditions did you find threatening, and what helpful? Try to relate them to what we will now discuss. We are going to cover the five steps in the relationship stage: (1) initiation/entry, (2) clarification of presenting problem, (3) structure/contract for the helping relationship, (4) intensive exploration of problems, and (5) establishment of possible goals and objectives of the helping relationship.

☐ Step 1: Initiation/Entry

A warm, smiling welcome is the best way to begin any interview or session. The obvious purpose is to put the helpee at ease and hence get down to the business of identifying issues and concerns as quickly as possible. And, of course, you want to let helpees know that you are genuinely glad to see them. Often some informal conversation about the weather, parking, and so forth is necessary to help the client relax. Ice-breaking remarks such as "Tell me what I can do for you" or "I'm interested in what's going on with you now" can help focus on the reason for the meeting.

Drawing Out the Helpee

At the beginning of any interview or session, you can better draw out the helpee and get more information by making responsive listening statements than by asking questions.

Some examples of drawing out the helpee follow.

Client: My husband has never been out of work so long before. It's having a terrible effect on the kids. That's why I'm here, I guess.

Helper: That's a scary situation for all of you, I know. Let's talk more about what it's doing to your family.

In the preceding example the helper is communicating support and understanding while at the same time steering the conversation toward specific effects of the husband's unemployment on his family.

Client: My friend Joan Astin said you helped her decide if she should stay with her boyfriend. I thought I'd come and see if you can help me decide what to do about my boyfriend. He just doesn't seem to want to make any decisions and I think he should know by now if he wants us to get married.

Helper: Joan is a delightful woman. I enjoyed working with her. You seem to be confused about what you want to do. Can you tell me more about your relationship and what you've been thinking about?

In this example, the helper is acknowledging the helpee's relationship with Ms. Astin and, at the same time, seeking clarification of the nature of the situation.

Client: *(fidgeting and avoiding eye contact)* I dunno why you sent for me. I ain't done nothin' wrong.

Helper: You feel uncomfortable because you think you wouldn't be here unless you were going to be landed on. I've asked you to come in so that I can get to know you better and find out more about how you're doing in shop. Mr. Jones seems to feel you're having some difficulties there.

Here the helper responds to the request for an explanation of the interview in a nonthreatening, honest manner.

Client: I don't know whether I should come to you with this. I hate to complain. My daughter is awfully unhappy with Betty [her camp counselor].

Helper: I'm glad you've come to see me, Mrs. Brown. Tell me what you think is troubling Sue about Betty.

In this case, the helper is assuring the helpee that it's O.K. to deal with this issue, while asking for some further information.

Friend: Hi! Got time for a cup of coffee?

Friend: Just a sec. I've been wondering how things are going with you.

In this informal setting the helping friend responds promptly to the cue for a visit and starts right in expressing concern.

Statements or leads such as those in the preceding examples, which draw out information in a nonthreatening, open, indirect manner, are door openers or ice breakers. Their purpose is to keep the communication flowing without any judging, confronting, or manipulating. Other such leads are "Tell me more about that," "I'm wondering about . . . ," "Seems to me that . . . ," and "That sounds really interesting." Minimal verbal statements can also effectively keep the communication flowing, as can head nodding, smiling, and affectionate shoulder pats.

Helpers need to be supportive and encouraging in order to establish an effective helping relationship. However, if you, as a helper, confuse reassurance with support and encouragement, you may aid people to avoid rather than approach their true concerns. If you are too reassuring, you are denying the legitimacy of the helpee's concerns and, by doing so, imposing your values and judgments, even though you think you are being nice and making someone else "feel better."

You may ask questions when you feel they will augment the communication flow, but it is best to make them indirect and open-ended. Most people ask too many questions to begin with, so it is a good idea for helpers-in-training to limit themselves to as few questions as possible and concentrate instead on responsive listening statements. You'd be surprised how often you can turn a question into a statement that will produce more information than the question would have.

Dealing with Reluctant Clients

The resistance of reluctant clients can negatively affect the counseling process and outcome. Helpers can learn to respectfully deal with this resistance from the outset. Larrabee (1982) suggests the use of affirmation techniques with reluctant clients. Based on responsive listening, these techniques help the client face the reality of his or her situation and understand that there are good reasons for maintaining that situation. For example, the helper might suggest to the client some of the possible outcomes of helping and propose a limited number of sessions to explore possibilities before making a final decision. The use of interchangeable and additive responses enables the helper to reflect the helpee's feelings and thoughts without communicating devaluation, sarcasm, or impatience—and at the same time allows the communication of genuine caring and respect for the helpee's different views. Larrabee also suggests that open-ended leads can help reluctant clients focus on self-examination. There are times, however, when it is necessary to use silence as the only way of responsibly attending to a reluctant client.

Exercise 4.1   Silence is a necessary technique for helpers but one they often feel uncomfortable with. Work through the following series of Gestalt-type exercises in triads, rotating the roles of helper, helpee, and observer each time after you've completed the series.

1. The helpee talks about a real or imagined concern, and the helper is not allowed to respond in any way, either verbally or nonverbally. After five minutes, share reactions and feelings.
2. Maintaining the same roles, repeat the first sequence; now the helper is allowed to make two nonverbal responses within a five-minute period. Again, process.
3. Now the helper can make one verbal and two nonverbal responses within a five-minute period. Process after completion.
4. The helper can make four responses, either verbally or nonverbally, within a five-minute period.
5. The helper can respond in any way and as often as he or she chooses in the last five-minute period.

Now discuss your reactions to the entire series of exercises, and then change roles and start over again.

Allowing sufficient time for the counseling process and exhibiting a great deal of patience and empathy also can help to diminish the resistance and defensiveness of reluctant clients. Your genuineness, which includes discussing alternatives and their consequences, enables the reluctant client to decide whether or not it will be in his or her best interests to cooperate with you. The client will not be able to make this decision, however, unless you can spell out just who you are, what you are doing there, and what you see as the nature and objectives of your prospective helping relationship.

Helpers often become enmeshed in a game with reluctant clients that does little more than pass the time. It is the "I'm only here to help you" kind of game, which is perceived as threatening by the client because it implies that the helper is omniscient. One way of combatting this game is to encourage the reluctant client to take the initiative for structuring the relationship. An example of this occurred in a halfway house, where the youth worker just stayed in close physical proximity to the reluctant client, shrugged off the verbal abuse, and communicated caring and genuineness more by nonverbal presence than by words. After four days of testing the youth worker in every conceivable fashion, the 14-year-old boy began to shed his bravado and share some of his real feelings and thoughts. It was almost a test of endurance: who could outlast whom.

Try not to feel guilty about or hurt by rejection from a reluctant client. Instead, try to become aware of possible aspects of your approach to this helpee, as well as aspects of your setting and situation, that may be contributing to that rejection. You may find that your style of helping is not working for a particular client in a particular setting. In that case, you can either modify your style or arrange for a referral. Modifying your style could include judicious use of humor or self-disclosure, or use of more open support and acceptance, or some temporary diversion, such as changing the subject. Other techniques include reflection of the resistance and clarification of the consequences of not working together.

☐ Step 2: Clarification of Presenting Problem

It may take some time and patience to uncover the problem that is of concern to the helpee. Understandably, most people test helpers with superficial concerns before they trust them enough to reveal more basic ones. For example, a woman came in to see me recently and told me she had come because she was worried about one of her children. However, as we proceeded to develop a relationship, it became apparent that she was angry at her husband and concerned about her marriage. In this case, it had been less painful for the woman to focus on her child than on her troubled marital relationship. You must actively listen and respond to the helpee in order to avoid being sidetracked by superficial concerns.

It is possible that a helpee will present several different concerns. By using responsive listening techniques, you can aid the helpee in sorting out and ranking the different problems.

The following examples show two different approaches to the same client statement expressing many concerns.

Client: I'm really having family problems. My fianceé and I have a mutual uncle and he is close to my parents and me. Her parents have not seen or spoken to him in 15 years. They have forbidden her to visit him or come to my house if he is there.

Helper: That's rough. Must make you feel guilty about your relations with your uncle.

Client: Yes. The problem is the wedding. My mother insists he come and my fianceé says it's impossible.

Helper: You're really in a bind and you don't know how to get around this.

Client: I'm beginning to think we should elope. You know large Italian family weddings; I don't know how we can make everyone happy.

Helper: Let's talk about the different ways you can deal with the wedding invitations.

.   .   .

Client: I'm really having family problems. My fianceé and I have a mutual uncle and he is close to my parents and me. Her parents have not seen or spoken to him in 15 years. They have forbidden her to visit him or come to my house if he is there.

Helper: You seem to be upset that your fianceé's parents have such control over her.

Client: Yes, I guess they really do.

Helper: It's uncomfortable for you to be in the middle of the two families, and for her not to be taking a stand with you.

Client: I think her first loyalty should be to me! After all, I'm going to be her husband!

Helper: Sounds to me like you're wondering how she'll react to her parents' pressures after you're married.

Client: That's really it.

The helper in the second example was able to clarify the real issue, rather than be sidetracked by an apparent problem. You can see how important it is to take the time to get to the real problem and not to rush too quickly into problem solving. It has been my experience, both as a counselor and with colleagues at meetings, that unless the time is taken to clearly identify the problem and its "ownership," a large amount of problem-solving activity is futile. What I mean by the phrase *problem ownership* is the identification of which person has the problem. One has to feel that one "owns" at least part of the problem before one will invest much energy in the problem-solving process.

Another example illustrating the problem-clarification step is the following:

Client: I'm not going to be able to come in for work tomorrow. I know you need me for the inventory, but I just can't make it.

Helper: Sounds like something is bothering you.

Client: No, not really. I'm going to have to start looking for a place of my own to live.

Helper: Your present living situation isn't working out.

Client: I'll say! My roommate is really bitchy and giving me a hard time. I need an apartment of my own, even though I don't know how I'll manage.

Helper: You seem scared about the financial responsibilities and torn between getting a place of your own and putting up with a bad situation.

Client: That's right. I've never been on my own before. I don't know if I can do it. Everything seems to cost so much.

Helper: Let's see if I can help you plot out what it will cost.

Again, by taking the time to listen to what the helpee was really saying, the helper was able to identify the helpee's underlying fear and insecurity, rather than be sidetracked by the helpee's relationship with the roommate.

In many cases, presenting problems cover up more pervasive underlying problems, and it may be many sessions before those problems emerge. This may be because some people take longer than others to develop trusting rapport or because the client does not understand or acknowledge to himself or herself what the real issues are. The steps of the relationship stage may be completed with the presenting problem and then begun again for newer, emerging problems.

☐ Step 3: Definition of Structure/Contract

Once the problem has been clarified and acknowledged by the helpee as one needing resolution, you can decide whether or not you are able to provide help in solving that kind of problem. If you feel that you are unable to provide help, you can aid the helpee in obtaining assistance elsewhere by means of **referral.**

Referrals are arrangements for clients to see a designated individual or agency for a specific purpose. In helping relationships, early referrals are especially important because they can take some time to effect. There may be a waiting period before a referral appointment can take place, and during that period you may want to maintain a supporting, encouraging relationship. The helpee may be reluctant to accept a referral, and acceptance (readiness) may become the goal for your helping relationship. If you serve in a helping role in a setting where personal or long-term counseling is not available, you can learn to identify available community resources so that you can quickly make contacts for helpees.

An example of a referral follows.

Ms. White, age 42, has been employed as a supervisor in the keypunch unit for eight years. She has recently lost a lot of weight and has displayed irritable, irrational behavior with her workers. Her behavior is unusual; for eight years she has been regarded as one of the most easygoing supervisors in the company. The following excerpt occurs at the end of the second session with the personnel manager, someone with whom Ms. White has talked before. Ms. White requested these sessions because she felt the need "to talk to someone."

Ms. White: And so it seems as if my whole world is falling apart. I can't eat, I can't sleep, I can't think.

Personnel Manager: I'm concerned about you. You're really having a difficult time.

Ms. White: I went to the doctor a couple of days ago. He says except for the weight loss and nerves that it's all in my head and I just have to stop worrying. I don't even know what I'm worrying about.

Personnel Manager: Thelma, I'd like to be able to help you, but I really think you need a different kind of help than I can give.

Ms. White: What do you mean? I'm not mental or anything like that.

Personnel Manager: No, I know that. But you are having problems that lots of people have at one time or another in their life. And there are people around who are trained to help you. They can help you find out what's worrying you.

Ms. White: I don't know. I don't think I could talk to anyone the way I can to you. And I can't afford much.

Personnel Manager: You can still talk to me. Let me try and find out who in your town can be of help, and the cost, and so forth. Then we can talk some more about it.

Ms. White: If you really think that's what we should do.

Personnel Manager: Can you come see me tomorrow morning? Around 9:45—just before coffee break? We can talk some more about it then.

Ms. White is frightened of a referral, and the personnel manager is sensitive to her fear. In a case such as this, the helper should take the time necessary to allow the trustful relationship with the helpee to serve as a vehicle for helping him or her to accept a referral. Because the personnel manager knows that Ms. White has already had a medical examination, she can take the time to be supportive and effect what she believes will be a satisfactory referral for Ms. White.

To feel good about making a referral, you must learn what you can and cannot deal with and learn who around you can handle what kinds of situations. However, if the nature and extent of the problem are such that you feel you can provide help, you must state clearly to the helpee just what you can and cannot do, what you expect from the helpee, how you perceive his or her expectations of you, and how much time you can devote to this helping relationship. These directions apply to both formal and informal helping relationships, whatever the setting. Unfortunately these guidelines are often neglected, and unhappiness and frustration result on both sides when (unclarified) expectations are unmet.

When the nature of the problem remains uncertain or other factors seem to be involved, you may want time to seek a consultant's advice about your own options as a helper, or you may suggest that the helpee seek consultation with another helper. If a medical condition, such as headaches or poor digestion, might possibly be contributing to a client's distress, you may want to refer the helpee to a physician before you proceed further. Or if a client is having problems with an ex-spouse, you may want to suggest legal consultation.

If you and the helpee agree to it, one way to clarify expectations is to make a contract. A contract can be written or spoken; it is clear, understood by all parties, and always open to revision. For example, if you tell a helpee you would like to meet five times and then see how things are, you can always decide to lengthen or shorten that time as long as you both agree. Agreement is the key factor. The terms of the contract can include time of sessions, length of sessions, site of sessions, fees (if applicable), an estimate of number of sessions needed, who may or will attend sessions, procedures for changing any of these terms, and identification of helper and helpee expectations.

The following example demonstrates the making of a contract.

Marianne, an 18-year-old college freshman, has been referred to the counselor by her residence advisor because of continuing homesickness. The following excerpt occurs at the end of the first session, during which the counselor has identified the problem as one of low self-concept with resulting feelings of inferiority and inadequacy.

Counselor: Our time is almost up, Marianne. I'm wondering how you feel about our talk today.

Marianne: It's been O.K.

Counselor: Would you like to come in again?

Marianne: Yes, I think so. Do you think I should?

Counselor: It's up to you. I do think we can talk some more about what you're doing and feeling.

Marianne: Uh-huh.

Counselor: Why don't we plan on having three more sessions—one hour per week—and then we can see where to go from there. You may want to think about joining a group then or you may want to continue coming in. We can leave that open for now.

Marianne: That sounds O.K. Do I make the appointments with you or the secretary out there?

In this case, the counselor suggests that there is more material to gather and that, in a few weeks, they may have some other options to consider. This kind of structuring provides a frame of reference so that the helpee does not feel caught up in an endless process.

Another example involves a reluctant client on probation after conviction on drug and truancy charges.

Scott, a 17-year-old high school junior, has been referred to a counselor by his probation officer. During the first session, he informs the counselor that his problems are due to his mother's homosexuality and that unless his mother is willing to change her sexual orientation, he cannot work through his issues. He is reluctant to return to the counselor and furious that the terms of his probation require weekly counseling sessions.

Counselor: I can understand that you don't want to come to see me.

Scott: I hate shrinks. I don't see why I have to come here.

Counselor: It's upsetting that you can't choose for yourself what you can do ... it must be hard to have someone else have control over you.

Scott: Yeah.

Counselor: I'm wondering what will happen if we decide not to meet again.

Scott: Make an appointment for me for next week.

Counselor: What? I'm surprised and confused ... thought you didn't want to.

Scott: Make the appointment. Mr. C. said I have to come for the rest of the school year, so I'll come six more times.

Counselor: Sounds like you want to finish school. I appreciate your frustration and we'll see what we can do to arrange for meetings the next six weeks.

In this case, the immediate problem was meeting the probation terms so that Scott could finish school. During the next six weeks, further clarification and exploration of presenting and underlying problems occurred. Note that the counselor did not fight the helpee's reluctance.

☐ Step 4: Intensive Exploration of Problems

Using the responsive listening model of communication, you can begin to aid the helpee to look at the aspects, implications, and ramifications of his or her problems. Of course, many new problems will emerge throughout the helping process. Again, it is important to know just what the problem is, who has responsibility for the problem, and to what extent the helpee is able to effect some problem solving.

A helpee's problem is often part of the system in which he or she lives. In such a case, the helpee is likely to feel very frustrated, hopeless, and helpless. However, the same principles apply in such cases as in other problem situations: you try to learn as much as you can about the helpee, the system, the possibilities for change, and the choices that exist. Throughout this period, you are learning more about the thinking processes, feelings, and behaviors of the helpee, both in and out of the helping relationship. You are discovering the client's values, beliefs, attitudes, defense and coping strategies, relationships with others, hopes, ambitions, and aspirations. This is a time when you encourage the helpee to express whatever thoughts or feelings he or she is experiencing, without fear of being judged or instructed. All the while, you are promoting the development of trust, genuineness, and empathy, so that you can create a safe climate in which the helpee feels free to explore his or her own self-awareness.

The following excerpt is from the third counseling session with Marianne and demonstrates intensive exploration of her problem.

Counselor: So you've felt for a long time that you couldn't do much on your own.

Marianne: My mom and Pat [older sister] always did everything for me.

Counselor: And that made you feel sort of . . .

Marianne: Dopey. They even would check over my homework every night. You see, Pat got married right after high school and never went on to college. She had been an honor student, and now here she is widowed and back home with a baby.

Counselor: Seems like you're supposed to make up to your mom and Pat for what they didn't have.

Marianne: They always say they want me to be and have what they didn't have. That's why it was so important for me to come to college.

Counselor: It was important to you, too?

Marianne: I don't know. I never thought about what was important to me. I never want to upset my mom; she cries so easily.

Counselor: And your dad . . .

Marianne: Oh, he wants whatever my mom wants. He never interferes or anything.

Counselor: I hear some anger in your voice.

Marianne: I was just thinking that there have been times when I wished my dad would stick up for me.

In this example, the counselor is learning as much as possible about Marianne's background and all the aspects of her feelings of inadequacy. This takes much time and skill, but some exploration is necessary before goals and objectives can be established. Intensive exploration is a continuing process that occurs throughout the helping relationship. For many helpers, it is one of the most challenging and exciting steps of the helping relationship. In fact, many relationships end at this step, because the act of mutual exploration is in itself therapeutic.

☐ Step 5: Establishment of Possible Goals and Objectives

After the problem has been thoroughly explored, the helper and helpee can more specifically develop goals and objectives for the relationship. This step can be accomplished in a systematic or casual fashion, depending on the style of the parties in the helping relationship. The important point is that both parties agree to whatever the goals and objectives are. It would certainly not be helpful for the client if the goals represented only the helper's needs.

Helpers and helpees will establish immediate and long-range goals, specific and diffuse goals. For example, a student who is referred to a counselor because he or she faces possible failure in a course may also have family problems. The counselor and the student may choose to work on the more specific, immediate goal—that of passing the course and staying in school—rather than the long-range, more diffuse goal of solving family problems. These goals may not be equally valid or important, so it is up to the helper and the helpee to determine mutually which goals are feasible given the nature and conditions of the particular helping relationship.

It is possible for the goal of the relationship to be simply to develop the relationship further, so that the helpee's self-concept is enhanced, or to provide a vehicle for self-understanding. That could very well become the goal in the case of Marianne. Or it might be that the goal is to make a decision or seek alternative forms of behavior. The point is that both the helper and the helpee need to know why their relationship exists and what their goals are. When I ask students what distinguishes a particular session in a counseling center from a friendly talk over coffee, they usually reply that a helping relationship involves goals and objectives that one does not consider in a friendly conversation.

When a helper and a helpee formulate several different goals, they should decide which goal has priority and how long that priority should last. Sometimes the ranking of goals falls into a logical sequence; other times helper and helpee arbitrarily decide on the order.

The following excerpt illustrates goal setting at the end of a first session.

Mr. Winsor, age 32, has come to see the adult advisor at the local community college. He wants to take some courses that will give him upward job mobility. He is feeling trapped in a job that will not take him anyplace.

Counselor: It's really important for you to see some possibilities of moving ahead.

Mr. Winsor: Yes. I have a growing family, we're having a hard time making ends meet, and we do want to be out of the city by the time the oldest starts school.

Counselor: But you're not really sure just what courses you might want to take or what kinds of jobs to aim for.

Mr. Winsor: At this point, I just want to move . . . up.

Counselor: It seems to me we ought to find out more about what kinds of work you'd really like and are suited for, rather than just sign you up for some courses.

Mr. Winsor: What do you mean?

Counselor: Why don't you take our battery of vocational interest and aptitude tests, and then we can go over the results and try to decide together
what would be the best path for you to follow?

Mr. Winsor: I guess so. Will it take long?

Counselor: I can set you up for a testing session next week and we can talk
before registration.

Mr. Winsor: I've heard about those tests . . . can't do any harm. I'm game.

The objectives of the relationship have become focused on vocational testing
and exploration.

Goal setting is also important with children. Like adults, children increase
their motivation—which usually results in better performance—by participating
in goal setting.

☐ Resistance

A discussion of the conditions and stages of a helping relationship would not be
complete without talking about resistance. Resistance can occur at any time in the
helping relationship. It typically surfaces when the helpee does not appear to
collaborate in developing the helping relationship or in establishing and/or
striving toward goals. Resistance is often the response when a helpee is feeling
threatened by the relationship or the material being explored or is feeling that the
helper is probing or interpreting sensitive issues before a trustful relationship has
been established.

Otani (1989) describes four categories for classifying aspects of client resistance: (1) amount of verbalization, (2) content of message, (3) style of communication, and (4) attitude toward helper and helping sessions. For example, clients
can show resistance by verbalizing too little, by limiting their verbalizing to safe
topics, by chattering irrelevantly or giving monosyllabic answers, by missing
sessions, being late, not paying attention, and so forth. Each of these categories
might have different meaning and relevance.

Most helpers find that a certain degree of resistance occurs in most relationships at some time. It can range from subtle forms of inattention, failure to keep
appointments, or other indications of ambivalent attitudes, to outright rejection of
the helper. The helpee's resistance guides the sensitive helper to focus more
closely on the characteristics of their relationship. In other words, one can be alert
to the helpee's resistance without reacting to it, while gaining an understanding
of his or her unique defensive style. The effective helper also tries to reduce
defensiveness, perhaps by changing the pace, topic, or level of the discussion, and
by communicating as much support and acceptance as possible. If the relationship

has already been well established when resistance occurs, the helper may perhaps reflect back the feelings of resistance and decide with the helpee how to deal with them. A helper can often reduce resistance by changing strategies or referring the helpee to a different source of help.

It is usually futile to get into a win/lose power struggle with a resistant client, as the helper can only lose. Remember, you can lead a horse to water but you cannot force it to drink! Struggling against the resistance only causes the helper to experience anxiety, frustration, and anger. The more a helper pursues a resistant helpee, the stronger and more manipulative the resistance will become. It is more useful to "go with" or follow the helpee's resistance in a supportive, nonthreatening manner. If all else fails and the helpee refuses to cooperate at all or attempt to deal with the resistance, a "sabbatical leave" from helping may be appropriate. Many helpees need time on their own to work through their resistances and, if a period of time is allowed between sessions, will feel more in control of the situation.

Often the resistance is the helpee's problem, but sometimes it results from inappropriate helper behavior. Blame is not the issue; what is important is understanding and recognition of the value and inevitability of resistance as part of the change process.

The following exercises will give you more practice in developing the communication skills that enhance helping relationships and uncover problems. They will also give you the opportunity to become familiar with the different steps of the relationship stage. The more you actively participate in and process exercises and discuss your feelings and reactions to them with others, the more meaningful the concepts discussed in each chapter will become.

Exercise 4.2   After each of the following helpee statements, circle the letter of the helper response that you believe would best facilitate the development of an empathic relationship and lead to clarification of the presenting problem.

1. **Helpee:** Ms. Smith said you wanted to see me.
   Helper:
   a. Yes, Mary. She tells me you're not getting on very well in that group.
   b. Oh? That's right, I did. Can you tell me what's going on in the steno pool? I understand production is way off.
   c. I have been wanting to talk to you, Mary. How do you feel about the way things are going in the steno pool?
   d. Yes. I want to see if there's any way we can help you feel good here. Can we talk about how things are going in the steno pool?
   e. Ummmm. I want to hear your version of what's going on in the steno pool.
2. **Helpee:** I'm not going to be able to get through that interview. I know I'll mess it up like all the other times!

Helper:

a. No, you won't—not if you make up your mind not to.

b. Come on! Have some faith in yourself. Of course you'll do O.K.

c. You're worried that you'll fall apart once you get in there?

d. Why don't you talk to some people who've already had their interviews and see if they can clue you in as to what it'll be like.

e. You feel that, because you've had some bad experiences interviewing, this one will be bad, too.

3. **Helpee:** I don't understand why we're always fighting. We just can't seem to talk about anything anymore.

   Helper:

   a. Tell me about it.

   b. All married people fight sometimes.

   c. It's frightening to be so angry with your husband most of the time. Makes you worry about what's happened to change your relationship.

   d. What do you fight about?

   e. You probably need a change. How about changing your routine a bit— maybe go out to dinner or to a movie or something that's different.

4. **Helpee:** I've come to see you because I need help. What do you think I should do about this?

   Helper:

   a. You seem to want me to tell you what to do.

   b. The first thing is to check out some information. Have you looked into financial aid?

   c. I can't solve your problems for you, you know.

   d. What do you want to do?

   e. I don't know. Let's talk some more about it.

5. **Helpee:** Business has really fallen off. They're talking about laying people off.

   Helper:

   a. You're worried that you may be laid off if things don't start to get better.

   b. I know. I'm worried too. Don't know what will happen.

   c. Yeah, times are rough. Every day the papers report more layoffs.

   d. Don't worry. I'm sure things are just exaggerated right now.

   e. Don't you think we're all in the same boat?

6. **Helpee:** I can't decide which high school to go to next year. I'm supposed to go to East, but I can get special permission to go out of the district to West.

   Helper:

   a. Sounds like you've been thinking a lot about it.

   b. What are the things you're looking for at each of the schools?

   c. Have you visited each of the schools yet?

   d. It's a hard decision to make. Let's talk about what you feel about each school.

e. I can see you're unhappy about having to make a decision. What could be done to help you?

7. **Helpee:** My coming to see you really won't help. You can't change anything.

Helper:

a. You wish I could make things easier for you.

b. You're pretty discouraged. You wish there were some way I could change things for you.

c. Sounds to me like you're not even willing to try.

d. Maybe there are some things *you* could change!

e. I'm wondering if you really want to change.

8. **Helpee:** If only I had listened to my dad. He told me this wouldn't work.

Helper:

a. Well, you live and learn. It's not so terrible to make a mistake.

b. Crying over spilt milk won't help.

c. It's rough. Tell me why you think it didn't work.

d. It's sort of scary to think your dad might be right about things. What else might he be right about?

e. It makes you nervous to find out you're not always right.

9. **Helpee:** I don't want to go back to school, ever. I'm not learning anything and the teachers are terrible. I want to get a job now.

Helper:

a. You know, John, if you don't finish school, you'll be sorry later on, and it will be too late.

b. I know how you feel, but it's pretty difficult getting a job these days.

c. You really are feeling down about that school. Sounds like you really want to get away from there.

d. You're not sure what you're getting out of being in school. Seems like a waste of time for you.

e. I'm wondering what you're doing to help yourself out over there.

10. **Helpee:** My parents are always on my back. They always want to know where I am, what I'm doing, and who I'm with. They never leave me alone.

Helper:

a. It does seem as if parents worry too much, doesn't it?

b. You're really upset because it seems to you that your parents don't trust you.

c. I'm sure they really love you and are just worried. It's so hard these days with teenagers.

d. Have you ever done anything to give them cause for worry?

e. It's nice to know they care so much, isn't it?

Remember, you do not want to come across as a judge, an interrogator, or a disinterested party. It is sometimes difficult for us to communicate warmth and acceptance because of our immediate reactions to and feelings about the helpee's statements and manner of presentation. Perhaps we've had a similar experience or problem, and we're anxious to pass on how we reacted and coped. There are times when such self-disclosure is facilitative, but in the initial steps of relationship building it is more likely to draw attention toward you and away from the helpee. Look at the suggested answers for this exercise at the end of the chapter, then discuss your answers in small groups, and see if you can determine why some are more appropriate than others.

Exercise 4.3   In small groups, discuss the following situations. Then, if possible, role play the courses of action listed and share with the other members of your group your feelings about and reactions to them. Give your rationale for accepting or rejecting each of the listed alternatives and then supply some of your own.

1. Mary, age 10, comes to you, her Girl Scout leader, to tell you that two other scouts in your troop stole some candy from the local dime store yesterday. She's been with them when they've done that before, too. What would you do, and why?
   a. Call in the other two girls and confront them with Mary's report.
   b. Discuss with Mary the morality of shoplifting, tattling, and one's association with wrongdoers even if one doesn't misbehave.
   c. Explore with Mary her own concerns about this problem and what her options are.

2. Mr. Green comes into your public employment office to insist that he did, in fact, show up for a scheduled job interview, even though the employer claims that he did not. What would you do, and why?
   a. Tell Mr. Green that you know he has lied and that you can no longer refer him for job interviews.
   b. Ask him to tell you what happened, and explain what you can and cannot do for him and what options are available to him.
   c. Suggest that he join a vocational counseling group for practice in interviewing.

3. Tom, age 14, comes to your gymnastics class at the local recreation center "stoned." You've been concerned about Tom's erratic participation (despite his steady attendance) in this class for a few weeks. Every time you've asked him what's wrong, he tells you, "Nothing. Don't bug me." What would you do, and why?
   a. Tell Tom that you want to have him in your class, but he really can't come when he is stoned, as that is not fair to you and the others in the group.
   b. Tell him that you're concerned about him and will be available to talk to him about whatever is bothering him whenever he's ready.
   c. Ask him if you can take him home and talk to his parents with him.

4. You are a store manager in a fashionable suburban shopping center and you have noticed for several weeks that your senior salesperson consistently has

been making errors during the nightly audit. You also have noticed that she is irritable and snapping at co-workers over every little thing.

   a. You tell her you've noticed she seems upset, and that you're concerned. Can she tell you what's going on?

   b. You explain to her that your regional manager is on your back about these errors, and you need to know if she needs help at night.

   c. You suggest to her that she seems to be troubled and perhaps needs to talk to someone. You know someone at a local agency that you would be happy to refer her to.

5. You are a volunteer for a local crafts program for the elderly. One of the gentlemen, Mr. Roberts, appears to be withdrawn and cranky. When you talk to him, he tells you that his daughter-in-law is being nasty to him, and he is unhappy living with her.

   a. You discuss with him how unsympathetic in-laws often are and sympathize with his position.

   b. You discuss with him the capabilities he has and the options he has for different activities within the community.

   c. You empathize with his loneliness and discuss what you and others can offer him in the way of companionship and activities.

Again, act out as many of the previous responses as you can, and see how effective they are. Get in touch with your own feelings and values as you identify with both helper and helpee.

Exercise 4.4   Which of the five steps of the relationship stage (initiation/ entry, clarification, definition of structure/contract, exploration, establishment of possible goals/objectives) is evident in the following example?

1. **Client:** If I am pregnant, my folks will murder me.
   Helper: You're scared of what your parents will do to you.
   Client: You don't know them! They always are yapping at me and telling me I'm gonna end up just like my sister.
   Helper: You don't want to be like your sister.
   Client: You bet your sweet ass I don't! She had to marry Ron when she was in high school, and now she's really stuck with two brats and a husband who beats her up.
   Helper: That's scary. You seem afraid that you may have gotten yourself into a situation where you can end up the same way.

In this excerpt, the counselor is in step 2: clarification of the problem. She is trying to determine whether possible pregnancy (the presenting symptom), relationship with parents, or fear of being "trapped" is the major issue.

Now decide which step each of the following examples illustrates, and discuss in pairs or small groups what you've decided. Then see the comments at the end of the chapter.

2. **Helper:** There are only two more weeks of school left. Would you like to spend that time working on your shyness in class, so that you can start off next year feeling more comfortable?
   Client: Do you think I'll ever be able to answer out loud in class without getting all messed up?
   Helper: There are exercises we can practice that will help you become less nervous.
   Client: That would really help. If I could learn that, my marks would go up.
3. **Client:** Am I too early?
   Helper: No, right on time. Come in and sit down.
   Client: Over here?
   Helper: Fine. Now, tell me about your father-in-law.
   Client: Yes, I'm so worried about him. You see . . .
4. **Client:** Do you think you can help me find a job?
   Helper: We can work together on preparing a resume and choosing leads to follow.
   Client: I don't really know where to start. What do you charge for this?
   Helper: Let's set up three sessions to get things rolling. Here's a copy of my fee schedule. There are also some forms for you to fill out.
5. **Client:** I just don't know whether it's best for me and everyone to stay with Joe or to leave.
   Helper: You feel a lot of ambivalence about your marriage.
   Client: Well, I don't want the children to grow up without a father, and he says we can't afford financially to split up. But then, I'm 37, and I hate to think of spending the rest of my life this way. There must be something better for me.
   Helper: Let's focus on this ambivalence for a while. What do you see yourself getting for yourself by staying in this marriage?

Exercise 4.5   In this exercise, if possible, you should form triads with people with whom you have not worked or formed relationships. Rotate the roles of helper, helpee, and observer, and select one or more of the following ten situations (or choose one of your own) as a presenting problem. Choose whatever setting feels comfortable to you. The helper is to work with the helpee for about half an hour; the observer uses the rating scales in Appendix A and makes notes about the behaviors of the helper and the steps he or she follows. At the end of the session, take as much time as necessary to fully process the exercise so that both the helpee and observer can provide as much specific feedback as possible to the helper. What steps were you able to accomplish as the helper? How did you feel? What do you feel you need to work on? This exercise is one that can be used for weeks of training, in or out of class.

1. You've had a blowup with your husband.
2. You're never able to get assignments in on time.
3. You want to lose weight.

4. You're afraid you'll be laid off at work.
5. Your car is in the shop for repairs, and you don't know how you'll manage, since you need a car for work.
6. You're afraid of flying, and you have an upcoming trip to make.
7. You're shy and have difficulty meeting new people.
8. You're not sure why you're here.
9. You have trouble saying no to people when they ask you for something.
10. You're worried about a friend who might have AIDS.

This exercise allows you to experience the power of the helper-helpee interaction. Helping relationships can vary in intensity, depending on the situation. Factors that affect the intensity are the helpee's expectations and feelings of insecurity and anxiety at the beginning of a session, as well as the expectations, feelings, competencies, and skills of the helper. Try to pinpoint these as you work through the exercises.

☐ Summary

In this chapter, we have discussed inherent conditions affecting the relationship stage. These conditions can be grouped under the categories of initial contact, duration of the helping relationship, applications and forms, physical facilities, timing of appointments, record keeping, and other people. An important distinction was made between interviews, initial meetings to establish rapport and determine whether or not the relationship is to proceed, and sessions, subsequent meetings that begin with the structure/contract step and last through termination of the strategies stage.

The five major steps of the relationship stage—initiation/entry, clarification of presenting problem, definition of structure/contract, intensive exploration of problems, and establishment of possible goals and objectives—were discussed, with illustrative case examples. An important point to remember is that each of these steps takes varying amounts of time and contact. In some cases, the first four steps occur in one or two meetings; in others, it takes many meetings before intensive exploration can occur. The amount of time spent and contact made in helping relationships is determined not only by the nature of the relationship and the issues addressed, but also by context. Some institutions and agencies have delineated the number of sessions possible; others are more flexible.

The issue of resistance was also presented in this chapter, with the viewpoint that supporting and understanding resistance are more effective than struggling against it. Resistance is a necessary part of the change and growth process, one that can be worked with rather than feared and avoided.

Research has been unable to prove that one helping strategy is more effective than another, but studies do show that the effect of any given helping relationship depends as much on the quality of the relationship as on the techniques and strategies used. With this knowledge, and with practice of the exercises in this chapter to help you gain proficiency in enhancing helping relationships, we can proceed to the next stage of the helping relationship: the strategy stage.

☐ Exercise Answers

Exercise 4.2   Possible answers: 1. d, 2. e, 3. c, 4. a, 5. a, 6. a or d, 7. b, 8. d or e, 9. c or d, 10. b

Exercise 4.4   Excerpt 2 represents the fifth step, setting possible goals and objectives. There is a time limitation here, and the helper and helpee choose to use the time focusing on assertive behaviors. Excerpt 3 represents the initiation/entry step. Excerpt 4 represents the structure/contract step, when fees, time, and specific commitments are discussed. Excerpt 5 is a brief sample of intensive exploration, when helper and helpee are covering all the aspects and consequences of the helpee's concerns.

☐ References and Further Reading

Benjamin, A. (1981). *The helping interview* (3rd ed.). Boston: Houghton Mifflin.

Brammer, L. (1983). *The helping relationship* (3rd ed.). Englewood Cliffs, NJ: Prentice-Hall.

Egan, G. (1990). *The skilled helper* (4th ed.). Pacific Grove, CA: Brooks/Cole.

Ivey, A. E. (1988). *Intentional interviewing and counseling: Facilitating client development.* Pacific Grove, CA: Brooks/Cole.

Larrabee, M. J. (1982). Working with reluctant clients through affirmation techniques. *Personnel and Guidance Journal, 61,* 105–109.

Otani, A. (1989). Client resistance in counseling: Its theoretical rationale and taxonomic classification. *Journal of Counseling and Development, 67,* 458–462.

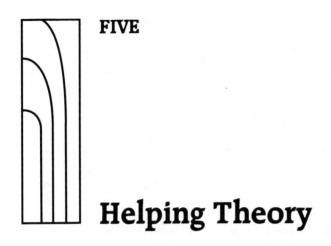

FIVE

Helping Theory

Now that we have covered the steps of the relationship stage, we can begin to examine the second stage of the helping relationship, the application of strategies. To provide a framework for understanding how to apply counseling skills and strategies, however, we must first review the major formal theories of helping. This chapter's review of those theories will cover their basic principles, their views of the helping relationship, their significant strategies, and their implications for helpers. Suggestions for further reading on each of the major theoretical viewpoints appear at the end of the chapter. In Chapters 6 and 7, we'll explore the strategic application of the major theories of helping.

Before studying these theories, we will explore the values and needs that comprise your "personal" theory. Understanding your personal theory of human behavior is important, because it will undoubtedly affect your understanding of, and your acceptance or rejection of, the formal, scientific theories of helping.

☐ Personal Theories of Human Behavior

You may now be asking yourself why we should bother exploring behavioral theories when we are learning about practical skills to apply in the context of unique helping situations. Perhaps you are asking this question because you are uncomfortable with what may seem to be an ivory tower approach to a down-to-earth situation. Furthermore, the term *theory* is somewhat threatening to many of us, possibly because it implies rigidity or the need to justify our positions.

However, let me point out that each of us already has personal views or assumptions that form a theory of use about human behavior. By *theory of use* I mean those beliefs and assumptions that operate in our day-to-day living. Our personal theories do exist and do affect our actions. Regardless of whether or not we are able to express them or are even aware of them, they affect our behavior, especially in interpersonal relationships. Our personal theories have been influenced by our ethnic, socioeconomic, and familial backgrounds; biological factors; gender; past experiences; exposure to different schools of thought; the people with whom we work and study; and opportunities. Our personalities and temperaments also influence our personal theories, as does our degree of self-awareness. Each of you will become more aware of your personal theory of helping as you think about it and as you explore helping situations.

Regardless of what our personal theory is, we need to understand its content because it affects our actions toward and reactions to the people we are helping. Unless we acknowledge our personal theoretical base, we may "help" people more in order to apply our theory than to satisfy their needs.

We should acknowledge that an ability to understand the concepts of human behavior is important for anyone working with people. The issues that we must grapple with as helpers—whether solely through the study of formal theories, through experiential learning, or through some integration of both types of learning—include how personality is formed, how personality is developed, why people behave as they do, what motivates people, how to motivate people, how people think, how people learn, if there are patterns in human development, what impact groups have on individual behavior, and how behavior can be changed.

We all have thoughts about these issues, although often we are not able to express our thoughts as theories. Nevertheless, our behavior and attitudes in interpersonal relationships are affected by our beliefs about these issues; and helpers who are aware of their beliefs and have consciously grappled with these questions are aware of the influence of their own theoretical views on their perceptions, attitudes, and behaviors toward helpees. For example, if we believe that by changing behavior we can change attitudes and feelings, we are more likely to adopt action-oriented counseling strategies focusing on behavioral changes in a formal counseling relationship than if we believe that behavior will change only through the development of self-awareness. In the latter case, we are more likely to adopt an approach using verbal techniques to develop insight. Likewise, an understanding of basic learning theory will enable us as helpers to be aware of and consciously use our potential as models in helping relationships.

Thus, knowledge of theory and its application is essential if we are to help effectively in a practical field. Each helper, however, develops a style of helping that is consistent with his or her personal theory. For example, how important to you is a personal relationship in helping, and how can you create the kind of helping relationship that is consistent with your personal theory? We can each answer this question differently and still be effective in our own way.

Unfortunately, there is often little congruence between what people involved in human behavior fields say they believe and what they actually practice. If we can become more aware of our own personal theory, we will be able to see how it relates to formal theory and to our own helping practices. To start thinking about some of your theoretical assumptions, ask yourself the following questions:

1. What are human beings? Are they good or bad? Are they born that way? Are they controlled or controlling? What motivates them?

2. What is my explanation of maleness and femaleness? Are differences due to biology? to socialization? Are racial and ethnic differences inherent or learned?
3. How do human beings learn? Are there different kinds of learning?
4. How do personality traits develop? Are they inherent or learned? Are certain personality types distinguishable by behavior?
5. Can people change? How do they change? Does something external cause them to change or does it come from within?
6. What is social deviance? Who decides what it is? What can or should be done about deviance? What behaviors (in myself and others) do I find acceptable and unacceptable? Are these behaviors deviant?

Try answering these questions in your own words, from your own frame of reference. You may be surprised to find that you do have answers to these questions! Compare them to others' answers. Your answers are neither right nor wrong, and today's answers could—and probably will—change as you continue to learn and become more aware of your own views and feelings.

How open are you to different viewpoints? Open-mindedness will affect your flexibility and adaptability, which in turn will affect the range of people with whom you're able to work and the settings in which you feel comfortable.

Now examine the following, necessarily brief, overview of the major theoretical views of helping, and see what parts of them you accept and reject. You may see some of your present assumptions reflected in these theories; however, your assumptions may change by the time you finish this book.

☐ Psychodynamic Theory

The psychodynamic approach to helping, which is based on psychoanalytic theory, was introduced by Freud and has the longest history of all the current theories. Freud is recognized for his great insight in developing procedures for treating disturbed persons based on his concepts and observations of mental and emotional processes. The development of many later theories of helping can ultimately be traced back to Freud's theory. Later psychoanalytic theorists include Alfred Adler, Erik Erikson, Erich Fromm, Karen Horney, Carl Jung, Wilhelm Reich, Harry Stack Sullivan, the object relations school, and self psychologists.

We'll now briefly review the Freudian contributions to psychodynamic theory. I want to stress, however, that today's psychodynamic approach ranges from orthodox Freudian to humanistic ego psychology and object relations. We'll also cover key concepts of Alfred Adler and Carl Jung, whose thinking was influenced by Freud's but developed into separate, major schools of psychodynamic theory.

Freudian Theory

Psychoanalysis and derivative psychoanalytic treatments comprise the strategies of Freudian theory. Freud's view of human beings, based on his clinical observations, was negative and pessimistic in that he perceived them as being inherently selfish, impulsive, and irrational. His view of human behavior was **deterministic**—that is, he saw behavior as predetermined by biological instincts and drives along with previous life experiences. This is a comprehensive view, encompassing people's inner experience, external behavior, biological nature, social roles, and individual and group functioning. Freud emphasized the goal of health as effective functioning in love, work, and play.

Freudian theory, based on a psychology of internal, intrapsychic conflict between competing internal drives or instincts, hypothesizes the existence of the following personality structures:

1. The **id,** an instinctually derived structure representing the primitive, selfish aspects of humans. The unconscious (unaware) id demands immediate gratification by increasing pleasure and reducing tensions.
2. The **ego,** a structure rationally attempting reality orientation by conscious mediating (thinking, perceiving, and deciding) among the aggressive id, the moralistic superego, and the external world.
3. The **superego,** a structure representing the internalization of parental, societal, and cultural moral and ethical injunctions (the conscience).

Behavior is considered to be the product of conflictual interaction among these three structures, which can occur consciously (with awareness) and/or unconsciously (without awareness). Freud's concept of the unconscious as a repository of all experiences and memories underlies his theory of motivation. According to Freudian theory, all psychological distress stems from the unconscious. **Anxiety,** the major symptom of distress, is caused by repressed emotions arising from basic conflicts among the id, the ego, and the superego over control of **libido.**

Libido is one of two motivational instincts postulated by Freud. The basic driving force of the personality, libido is stored in the id. It comprises survival or life instincts such as hunger and sex. Aggression, the second motivational instinct, comprises death instincts such as hostility and self-destruction. Thus, libido is the drive energy directed at objects needed to fulfill instincts and biological needs. Libido must be either diverted or discharged.

Neuroses are psychological disorders that occur when the superego impairs ego functioning by imposing guilt, in order to limit the impulses of the unconscious id. They may be experienced as somatic and/or behavioral symptoms, such as depression or compulsive eating. When conflict between impulses and wishes is suppressed, anxiety results that triggers the ego **defense mechanisms** (such as **repression, isolation, regression, rationalization, reaction-formation, sublimation, projection, denial, compensation,** and **intellectualization**). The purpose of the defense mechanisms is to reduce anxiety. While successfully

reducing anxiety, defense mechanisms can obscure the true nature of the intrapsychic conflict and, therefore, can result in neuroses. Neuroses are derived from
childhood conflicts and can emerge in childhood, adolescence, or adulthood,
when the balance between the libidinal drives and the ego defense mechanisms
is upset and anxiety is heightened. This lack of balance can occur during normal
developmental transitions, or through disappointment, loss of love, physical
illness, or other crises or situations that trigger repressed childhood material to
spill over into current functioning.

Freud also conceptualized five **psychosexual stages** from birth through
adolescence for the development of libido. These stages have a crucial effect on
personality development, in that a healthy personality cannot develop without
their successful resolution. They are as follows:

1. The **oral stage,** the first year or so of life, when the infant receives pleasure
   from sucking and biting and develops trust. This is a pleasurable stage and
   we find that people whose oral needs were gratified tend to have a more
   positive view of the world in later life.
2. The **anal stage,** around the second and third years of life, when the child
   receives pleasure from stimulation of the anal erogenous zone and experiences the power of retention and excretion, as well as the negative
   feelings that accompany that power. The child shows a great deal of interest
   in bodily processes, and in smelling, touching, and playing with feces.
   Because of the power struggles resulting from parental attempts to toilet
   train, the child exhibits stubborn, rebellious behaviors. People whose anal
   needs were frustrated tend to become stubborn and retentive, with a
   strong need to control.
3. The **phallic stage,** the next few years, when the child receives pleasure
   from the stimulation of the genital region and begins to develop love-
   hate relationships with others. During this period oedipal longings
   occur: intense, erotic desires for the parent of the opposite sex and
   hostile, competitive feelings toward the parent of the same sex (the
   Electra complex or the **Oedipal complex**). Girls may manifest **penis
   envy** during this stage, as boys gain more privileges through their
   identification with their apparently "freer, more powerful" father.
   Freud believed that later neurosis involved the repressed wishes of the
   oedipal period, and that this stage laid the foundation for adult sexual
   relationships.
4. The **latency stage,** from around 6 to 12 years, when the child appears to
   lose interest in sexual fantasies and genital stimulation. This is a quiet
   phase, when the child focuses on socialization, exhibiting most of his or
   her energy in school and the outside world.
5. The **genital stage,** which occurs at puberty, when the adolescent begins
   to focus his or her concerns and gratifications on others rather than on the
   self and begins to develop heterosexual relationships.

According to the Freudian school, neuroses can be traced to a fixation at any one of the psychosexual stages. Thus, adult problems involving distrust in relationships, low self-esteem, inability to recognize and express negative feelings, and inability to accept one's own sexuality and sexual feelings can result from impeded development in one of these psychosexual stages.

The following are the major Freudian psychoanalytic constructs:

1. Human behavior is determined by unconscious forces (biological and instinctual needs and drives).
2. Sex drives are the principal determinants of behavior and underlie the dynamic conflicts, such as competition and jealousy, arising from early childhood.
3. Adult behavior is greatly influenced by early childhood experiences of the psychosexual developmental stages.
4. There is a fixed quantity of libido stored in the id.
5. Problems arise from intrapsychic conflict represented by anxiety and caused by past occurrences. This conflict results in the excessive use of defense mechanisms, which uses up an inordinate amount of libido. Thus, little libido is available for effective functioning in current life.

Jungian Theory

Other psychodynamic views are not as negative as Freud's. Carl Jung focused on the role of *purpose* in human development, presenting a more creative, optimistic view of humankind. He believed that one had more energy than that derived from sexual drives and that one was always developing toward wholeness and self-fulfillment (individuation), using energy to be "creatively purposeful" and searching for a balance among body, mind, and spirit. Jung believed that "transcendent function" mediates the relations between the conscious and the unconscious. He differentiated between the personal unconscious (painful, threatening experiences repressed or ignored) and the collective unconscious (buried memories based on the wisdom of the ancestral past). This differentiation helps in understanding and interpreting unconscious material—that is, symbols. Through his construct of the collective unconscious, Jung paid more attention than did Freud to the role of culture in the development of the human personality. In time of war, for example, one can imagine the forces of the collective unconscious causing people to unite in aggression.

Other major Jungian concepts include the **persona** (public mask or social facade one displays in various situations), the **animus** (masculine side), the **anima** (feminine side), **extroversion** (orientation toward the outer, objective world), and **introversion** (orientation toward the inner, subjective world). These concepts led to Jung's postulation of four types of people: thinking, feeling, sensing, and intuiting. In distinguishing among these psychological types, Jung pointed out that different types communicate with great difficulty. For example, an intuitive person will be impatient with the workaday practicality of the sensation-oriented person. The thinking type will have trouble with the feeling type, and vice versa. All these types, in varying degrees of extroversion and introversion, exist within every human; however, one type tends to be more pronounced than the others.

Adlerian Theory

Alfred Adler, like Carl Jung, differed with Freud's biological, deterministic viewpoint. He viewed humans from a nondeterministic, social-psychological perspective. He emphasized the social determinants of personality and focused more on the conscious than on the unconscious, more on the future than on the past, and on people's power to control their destinies. Contrary to Freud or Jung, Adler believed that people have social, creative, decision-making capacities with which to reach their selected life goals.

Adlerian personality theory includes the following concepts:

1. *Teleology:* Behavior is purposeful and goal-oriented and consciously selected.
2. *Inherent inferiority:* Humans strive for perfection, significance, and superiority to compensate for their basic **inherent inferiority** feelings.
3. *Compensation:* While striving for power to overcome inferiority, people attempt to compensate—to translate weakness into strength or find a particular area of competence that will overshadow weakness.
4. *Lifestyle:* Each individual develops a unique **lifestyle** early in childhood as a way of compensating for inferiority and weakness.
5. *Birth order:* One's **birth order** (status) in one's family of origin strongly influences one's childhood experiences and the development of lifestyle.

In Adlerian theory, maladaptive behavior occurs when an individual develops inappropriate strategies to overcompensate for the feelings of inferiority that emerged during childhood.

Ego Psychology

In the mid-1900s, ego psychologists (such as Anna Freud, Heinz Hartmann, and Erik Erikson) differed from Freud by paying more attention to the influence of the environment on personality development and to the ego's capacity to transcend the instincts of the id. Erikson proposed eight stages of development, which extend Freud's psychosexual stages throughout the life span and include psychosocial dimensions. Each stage provides the opportunity to resolve a core crisis.

1. *Infancy:* trust versus mistrust based on parent-infant relationship
2. *Early childhood:* autonomy versus shame and doubt based on separation from parent and independence experiences
3. *Preschool age:* initiative versus guilt based on initiative and competence opportunities

4. *School age:* industry versus inferiority based on learning skills and achievement experiences
5. *Adolescence:* identity versus role confusion based on individuation and lifestyle choice opportunities
6. *Young adulthood:* intimacy versus isolation based on peer relationship experiences
7. *Middle age:* generativity versus stagnation based on helping younger people
8. *Later life:* integrity versus despair based on reasonable satisfaction with one's life experiences

Object Relations and Self Psychology

Contemporary psychoanalytic thinkers (primarily Melanie Klein, Ronald Fairbairn, D. W. Winnicott, Harry Guntrip, Margaret Mahler, Heinz Kohut, and Otto Kernberg) emphasize the ego and interpersonal relationships rather than the id and innate biological drives as the basis of personality development. A person's primary drive from birth is considered to be for relationship contact, as opposed to sexual drives. The focus is on the relationship between an individual and real people, between an individual and his or her mental image or representation of real people, and between an individual's mental images or representations from early significant relationships and current real people.

The crucial development of personality begins in the pre-oedipal phase, with the infant's relationship with the primary caregiver, usually the mother, rather than with the infant's relationship with the father during the Freudian oedipal phase. Mother becomes the infant's first love object. In the first stages of infancy, when the infant cannot differentiate self from other, the infant ego's ability to develop a sense of security within itself and the environment depends on the identification the mother feels with her infant and on her capacity to empathize and nurture. If the infant's needs are not met during this symbiotic phase, the infant's ego splits, withdrawing and hiding to avoid the anxiety resulting from not having primary essential needs consistently met. The ego splits into what Winnicott (1965) calls the true self and the false self. The true self is at the core of human existence and is able to relate to itself and to others. The false self arises as a protection for an undernourished, insecure ego. It hides from the outer world and relationships. Thus, shortcomings or failures in early maternal nurturance lead to false selves and inhibit the development of a whole ego. Aggression is viewed by these theorists as a response or reaction to frustrating relationships rather than as an instinct.

The ego passes through many stages during infancy and early childhood, from the symbiotic relationship with the mother on through the separation and individuation stages. The quality of the infant's experience with object relations in early years shapes the development of the ego, which includes the capacity to love and relate to others. **Splitting** because of poor object relations disturbs this developmental process and can contribute to such pathologies as narcissistic character disorders, borderline states, and psychoses.

Psychodynamic Theory's Major Principles of Helping

The purpose of psychoanalytic treatment is to make the client conscious of unconscious material and to restructure his or her personality in order to attain a healthy balance of energy. For the fulfillment of this purpose, Jungians stress the need for a restructuring of basic character, while Adlerians stress the need to change the client's self-concept and restructure his or her development, lifestyle, and social situation. Ego psychology and object relations practitioners stress the nurturing quality of the therapeutic relationship, to provide the object constancy lacking in early infancy and to allow the ego to reintegrate and develop.

Psychoanalysis is a moderately **directive therapy** within an interview format. It uses such techniques as questioning, interpretation, dream analysis, **free association** (stream of consciousness), **recall,** and the analysis of **resistance, transference,** and **countertransference.** Some of these techniques (for example, questioning and interpretation) are directive whereas others (for example, free association and recall) are nondirective.

The four major phases of psychodynamic psychotherapy are (1) the opening phase, (2) the development of transference, (3) the working-through of transference, and (4) the resolution of transference. Transference is an integral part of the helping relationship. It occurs when the client revives emotions and attitudes originally present in the parent-child relationship and directs them toward the therapist. For example, a client may see the therapist as his father with his father's attributes and feelings, and may experience powerful feelings toward the therapist based on that perception. The client may be able to work through his unresolved conflicts with his father by means of this transference relationship. Further, analysis of a transference can help a client understand how he or she misperceives, misinterprets, and misresponds to people in the present in terms of past relationships.

The working-through phase involves interpretation of material generated by the client through free association, reporting of dreams, and recall. The goal is to help the client achieve insight and a corrective emotional experience by means of **abreaction** (the expression and discharge of repressed emotions via catharsis).

In psychoanalysis, the client is encouraged to experience crises, resistance, and transference in order to work through impasses and unconscious material. Psychoanalysts and psychodynamically oriented therapists stress the past (attempting to connect the past to the present), the exploration of causality, and the confrontation of client discrepancies.

Countertransference occurs when the therapist develops feelings toward or views about the client that stem from the therapist's conflicts rather than the client's. Countertransference can take the form of positive (love or excessive attachment) or negative (dislike or hostility) feelings and result in distorted interpretations. However, the psychoanalyst is trained to be objective and to work through his or her own conflicts in order to avoid personalizing feelings projected by the client.

Classical psychoanalytic theory postulates a directive, authoritarian relationship between the helper and the helpee. The relationship is not reciprocal; the helper is an expert authority figure and remains detached, objective, and completely neutral, so that transference will develop without contamination by the therapist's personhood. However, Adler introduced the concept of empathy into the psychoanalytic helping relationship, and today, buttressed by the focus on empathic relationships by object relations theorists, therapists of this school are encouraged to demonstrate natural warmth and empathy as important components of the helping relationship.

Implications of the Psychodynamic Approach for Helpers

To practice this type of therapy, the helper must undergo many years of rigorous training, so this approach is not directly relevant for counselors and helpers in other than traditional clinical settings. However, it does have important implications for nonpsychoanalytic helpers. Helpers need to recognize that there are motivating forces within people that are not wholly conscious; they need to know how people defend themselves from internal and external threats with defense mechanisms and resistance; and they must understand the significance of early childhood experiences and the possible implications of the concept of transference for any relationship.

☐ Phenomenological Theory

The phenomenological theories of helping focus on the uniqueness of each person's internal perspective, which determines one's reality. This approach emphasizes the here and now rather than what was or what will be, and how people perceive and feel about themselves and their environment rather than their adjustment to prevailing cultural norms. It also emphasizes affective rather than cognitive or behavioral domains.

The three most widely used phenomenological approaches are (1) existential psychotherapy, expounded on in this country by Rollo May, Viktor Frankl, James Bugental, and Irvin Yalom; (2) client-centered (person-centered) theory, developed by Carl Rogers; and (3) the Gestalt theory advanced by Fritz Perls. While we will focus more on the theories of Rogers and Perls, a brief description of the existential approach provides a glimpse of the philosophical framework that underlies these theories.

Existential Theory

This philosophical orientation stems from the thinking of nineteenth-century European philosophers and was developed into an approach to therapy by European analysts in reaction to the determinism of psychoanalysis and behaviorism. Thus, the existential view of human nature is that it is subjective and individualistic, that meaning is whatever one uniquely experiences. Human beings are always in the process of becoming, and they have the capacity for awareness and the freedom and responsibility to make choices. The goal of therapy is to enable clients to recognize the full range of their choices and to take the responsibility for whichever option they select. Anxiety, an inevitable part of the human condition, emanates from the awareness of death, freedom, isolation, and meaninglessness, and can lead to excessive use of defense mechanisms and lack of authenticity. The existential therapist serves as a model and a companion in the client's search for awareness, responsibility, and meaning. Thus, the nature and philosophy of the helping relationship are the most critical therapeutic variables. Carl Rogers elaborates upon the ingredients of this crucial person-to-person relationship.

Client-Centered (Person-Centered) Theory

The client-centered approach, established during the 1930s and 1940s by Carl Rogers, was largely a reaction to the rigidity of the psychoanalytic school that dominated the helping professions in the United States during that period.

Client-Centered Theory's Major Principles of Helping

Contrary to the psychoanalytic view, client-centered theory assumes that human beings are rational, good, and capable of assuming responsibility for themselves and making the choices that can lead to independence, self-actualization, and autonomy. Further, it proposes that people are constructive, cooperative, trustworthy, realistic, and social. The theory does admit that negative emotions such as hate and anger exist, but it holds that they exist mainly as responses to frustrated needs for love, security, and a feeling of belonging, which are basic human needs. This is a "self" theory, based on a belief that people act in accordance with their **self-concept** and that their self-concept is heavily influenced by their experiences with others. It is phenomenological in that it is concerned with the client's perception of his or her self and situation, not the helper's or the outside world's perception of the client. This theory is not concerned with causes of behavior or with changing behavior; rather, it focuses on the individual's current experiences, feelings, and interactions.

This theory emphasizes, then, the self, the environment, and the interaction of the two. The aware, self-actualizing organism is constantly experiencing in a perceptual field. The part of that field that one accepts or experiences as separate from the rest becomes the self. Since the self, by nature, strives toward integration and actualization of potential, the self-concept becomes increasingly harmonious and increasingly consistent with experience, continually accepting and integrating experience as a part of the self-structure. An individual's experiences, feelings, and interactions may either be integrated into the self from the environment to become part of the self-concept or remain part of the environment, not yet integrated into the self.

The following are the major client-centered constructs:

1. Self-concept comprises the individual's perceptions of himself or herself based on interactions with others.
2. The phenomenal field is the individual's reality and consists of his or her self-concept and perceptions of his or her world.
3. Individuals behave in whatever ways will enhance their self-concept.
4. Problems arise out of incongruencies between the individual's self-concept and life experiences that become threatening and cause the individual to use defenses such as the denial or distortion of experiences. These incongruencies lead to disorganization and pain.
5. Only by receiving unconditional positive regard (acceptance) from a significant other can persons be open to their experiences and develop more congruence between self-concept and behavior.

The foundation of client-centered therapy is the creation of an empathic relationship between therapist and client that will allow the client to experience spontaneity, genuineness, and "here-and-now" feelings. This principle applies to clients ranging from "normal" to "severely disturbed." The goals of this therapy are self-actualization and complete self-realization, and these goals can be achieved if the counselor can understand and empathize with the unique experiential world of the client. The counseling relationship creates a climate conducive to the client's self-exploration and gradual openness to change and growth. This type of therapy requires full participation on the part of both therapist and client. The therapist uses himself or herself as a vehicle of change by facilitating a growth-oriented, client-centered relationship.

Major verbal techniques employed by the client-centered helper are minimal leads such as "Mm-hmm," "I see," and "Yes" (connoting acceptance); reflection (a verbal statement mirroring the client's statement); clarification (which explores and develops a client's statement); summarization (which synthesizes a number of the client's statements); and confrontation (a verbal statement that non-judgmentally encounters a client's statement). Note that the client does all of the leading and directing, while the counselor follows the client's lead. Thus, there are no "techniques" in this form of helping. The helper is effective by being genuine and not playing the role of a "helper."

Carl Rogers's early work (1940–1950) mainly advocated use of nondirective "parroting," or "paraphrasing," of client statements in a permissive environment. He saw the therapist as a clarifier who should remain fairly distant and apart. Later (1950–1957), Rogers began to advocate a degree of sharing through self-disclosure on the part of the counselor, and some interpretation of feelings. More recently, Rogers became concerned with further increasing the counselor's active involvement with clients through self-disclosure and the sharing of counselor attitudes. Thus, Rogers came to advocate that counselors experience themselves as persons in relation to clients. Rogerian counselors today are more active and involved with their clients than were early Rogerian helpers, and they have added questions and feedback to their verbal response repertoires.

More than any other helping approach, the client-centered theory focuses on the relationship between helper and helpee, which is nondirective and emphasizes the communication of respect, understanding, and acceptance by the helper. The goal is an affective, warm relationship that reduces the helpee's anxiety and frees the helpee for experiencing, expressing, and exploring his or her feelings. The helper is presented as an equal, a co-worker of the helpee, not an expert or authority. And the client experiences the helping relationship as one that allows him or her to take responsibility for determining goals and for taking action toward those goals.

Implications of the Client-Centered Approach for Helpers

Because client-centered therapy emphasizes relationships based more on empathic and accurate listening and openness than on techniques, the counselor's self or "being" (attitude) is seen as more important than his or her acts or "doing." Thus, the helper is with the helpee on the helpee's level and in the helpee's frame of reference. The helpee is encouraged to continue to develop his or her self-awareness at the same time that the helper continues to develop his or her self-awareness and ability to elicit feelings in order to facilitate growth and responsibility in clients. This focus on the process rather than the content of verbal behavior is a major contribution to counseling psychology, and it has provided the foundation for counselor training programs and research.

Gestalt Theory

The Gestalt approach to helping is based on the perceptual learning theory developed by Kurt Koffka, Wolfgang Köhler, and Max Wertheimer, although the application of this theory as therapy was developed by Fritz Perls in the late 1940s. *Gestalt* is a German word meaning "configuration." All human behaviors and experiences are organized into Gestalts, into configurations or patterns, in which the whole is greater than the sum of its parts. Individuals form meaningful wholes (patterns) out of their experiences, and their needs determine whether parts of their experience (events) become dominant or background material. Likewise, individuals and their behaviors must be perceived as wholes. The organism is contained within his or her environment by an ego boundary. The environment is the source of activities, people, and experiences to fill the individual's needs. It is up to the self-aware individual to take responsibility for seeking from the environment that which will meet his or her need to become more self-supportive and psychologically stable.

Like client-centered therapy, this approach is phenomenological in that it focuses on the present, the here and now, and the client's perspective, rather than on the problem's original cause. Its concern is with increasing understanding and emotional and physical awareness of and by the client, which leads to integration of the parts of self and experience—eliminating any discrepancies—into a unified whole. Thus, Gestalt therapy is experiential (it emphasizes doing and acting out, not just talking), existential (it helps people to make independent choices and be responsible), and experimental (it encourages trying out new expressions of feelings).

Gestalt Theory's Major Principles of Helping

Gestalt theory holds that one can be responsible for one's actions and experiences and can live as a fully integrated, effectively functioning individual. The difficulty in assuming this responsibility results from a developmental impasse in one's past, a block that keeps one from living fully in the present and understanding the *how* and *what* (not the *why*) of behavior. Impasses, or inconsistencies, between the organism and its environment create conflict—that is, avoidance of contact and denying, negating, covering up a present experience rather than accepting it, and emphasizing what is not present rather than what is present.

Gestalt theory emphasizes the whole person (mind and body are seen as the same, not separate), and it regards the self as a total organism as it responds to the environment. Only the *how* of the present is considered legitimate material for discussion, not the *why* and *when,* the past, or the future. One can reminisce about the past and daydream about the future, but one must live in the present to be a fully functioning individual. The individual must develop self-awareness, acceptance, wholeness, and responsibility in order to achieve **organismic** balance. Feelings are considered to be energy. A person gets into trouble by not expressing feelings and by accumulating **unfinished business,** which results in tension and somatic (physical) difficulties and does not allow problems to be solved.

The major constructs of Gestalt theory are the following:

1. Maturity (wholeness) is achieved when persons are able to be self-supportive rather than environment-supported, when they are able to mobilize and use their own resources rather than manipulate others, and when they are able to accept responsibility for their own behaviors and experiences.
2. Awareness reduces avoidance behavior by allowing one to face and accept previously denied parts of one's being in order to become whole.
3. Change occurs when people assume responsibility for themselves and when they terminate unfinished business (which usually consists of unexpressed feelings related to past events that currently interfere with one's functioning, thus preventing one from acting in the present).
4. Focus of therapy is on the individual's current feelings and thoughts; on exploring all of his or her sensations, fantasies, perceptions, and dreams; and on encouraging him or her to take responsibility for them and "own" them in order to achieve integration.

5. The individual is encouraged to trust his or her intuitive sense rather than adjust to society.

Gestalt therapy uses a workshop approach to helping, with one-to-one relations being developed within a group setting, although the techniques can be applied on an individual basis. Major techniques of Gestalt therapy include the use of rules and games. Their goal is to interrelate mental activity, feelings, bodily sensations, and actions. Body language and an awareness of it are as important as the awareness of verbal language.

In Gestalt therapy, the therapist acts as a catalyst in that he or she directs, challenges, and frustrates clients so they may develop awareness of their whole (Gestalt). As previously mentioned, vehicles for expression include acting out exercises and games—that is, acting out conflicts in the present through exaggeration and role reversals. When Gestalt therapists ask clients to engage in certain role-playing exercises, they direct the clients to focus on certain details and feelings in order to push for client responsibility. The Gestalt therapist refuses to allow clients to avoid present experiences by intellectualizing ("talking about"), escaping to the past, or daydreaming about the future. The therapist's interventions involve asking "what" and "how" questions (never "why"!) to help expand the individual's sense of responsibility (owning the problem) and awareness and make the implicit explicit through the exaggeration of behavior.

Like client-centered therapy, Gestalt therapy focuses on the individual and an authentic helping relationship in the here and now rather than on interpretation. Unlike client-centered therapy, the Gestalt approach is directive in that the therapist is like the director of a play.

Implications of the Gestalt Approach for Helpers
In Gestalt therapy, the therapist is involved in a learning situation with clients and has the skill to teach them how to learn about themselves and become aware of how they function. The focus in Gestalt therapy is on both verbal and nonverbal techniques. Warmth and empathy are not emphasized as in client-centered helping relationships, but unless the client has faith in the therapist's **potency** (skill and ability to help), it is unlikely that Gestalt therapy will be effective, because the client will resist the therapist's suggestions and directions to carry out specific exercises.

The Gestalt therapist is confrontational and frustrating. For example, the therapist often asks the client to "stay with this feeling," which is frustrating for the client who wants to avoid that particular feeling. In this way, the Gestalt therapist encourages the client to analyze and reintegrate the unpleasant feeling with his or her whole self. The Gestalt therapist attends to the client's body language and looks for incongruencies in awareness and attention.

There are several Gestalt principles that helpers without extensive training in Gestalt therapy can use. For instance, they can point out clients' body (nonverbal) language as well as verbal language and attempt to integrate these into a whole. They can also emphasize clients' responsibility for their feelings, thoughts, and actions in the here and now; use techniques to encourage clients' self-awareness and self-reliance; and recognize the effect of "unfinished business" on current functioning. Gestalt therapy deemphasizes abstract intellectualization and provides a dramatic, quick methodology for enhancing self-awareness.

☐ Behavioral Theory

Unlike the psychodynamic and phenomenological approaches, which were developed from clinical practice, the behavioral approaches to helping were developed in psychological laboratories. The ethos of behavioral approaches arose from the inability of scientists to measure and evaluate the outcomes of psychoanalytic and phenomenological approaches to helping and from a need to predict and measure outcomes of helping based on specific, observable, objective, and measurable variables (that is, overt cognitive, motor, and emotional behaviors). The reasoning went that we cannot see, therefore we cannot measure, feelings or thoughts, so we must concern ourselves only with behavior in order to be truly accountable as helpers.

In the 1960s, behavior therapy was considered the clinical application of classical and operant conditioning theory. Today it comprises various approaches, such as applied behavior analysis, stimulus-response approaches, behavior modification, and social learning theory. For the purposes of this text, we will focus on the basic, common concepts that distinguish behavioral theory from other helping theories.

Behavioral Theory's Major Principles of Helping

According to behavioral theory, human behavior is determined by its immediate consequences in the environment (**reinforcement**). Therefore, it is learned from situational factors, not from within the organism. All behavior is learned and, thus, can be unlearned. Difficulties occur when learned, maladaptive behavior results in anxiety; the anxiety is learned as a contingency of the learned maladaptive behavior. This view assumes that people have no internal control over their behavior, no self-determinism: all behavior is determined by environmental variables. The human being is viewed as an organism capable of being manipulated. Values, feelings, and thoughts are ignored; only concrete, observable behaviors are considered. As a result, all behavioral theories are committed to a rigorous scientific method, with emphasis on behavioral analysis and treatment evaluation.

The major constructs of this theory are the following:

1. All behavior is caused by the environment (stimuli).
2. Behavior is **shaped** (the **principle of gradation**) and maintained by its consequences (responses).
3. Behavior is determined by immediate rather than historical antecedents.

4. Behavior that is reinforced by either **concrete reinforcement** or **social reinforcement** is more likely to recur than behavior that is not reinforced.
5. Positive reinforcement has more conditioning potency than negative reinforcement.
6. Reinforcement must follow immediately after the behavior has occurred.
7. Reinforcement may be either concrete or social.
8. Behavior can be extinguished by the absence of reinforcement.
9. Behaviors can be shaped by reinforcing successive approximations of the desired behavior.

The behavioral helping approaches are specifically directive and controlled; they rely on learning processes and cognitive mechanisms. The helper identifies unsuitable stimulus-response bonds (causes and effects of the target behaviors) and arranges to interfere with or to extinguish these unsuitable bonds. The helper then sets up conditions for teaching new, more desirable stimulus-response bonds so that more appropriate behaviors will be learned. These principles of reinforcement allow for both **discrimination** among stimuli and **generalization** of learning from one situation to another. The four general behavior modification approaches are (1) imitative learning (modeling), which teaches new behaviors by actual or simulated (video or audio tape) performing of desired behaviors by models; (2) cognitive learning, which teaches new behavior by role playing, rehearsals, verbal instructions, or **contingency contracts** between helper and helpee that spell out clearly what the helpee is to do and what the consequences or reinforcement of this behavior will be (a **token economy**); (3) emotional learning, such as **implosive therapy** (massive, exaggerated exposure to highly unpleasant stimuli in imaginal form to extinguish associated anxiety), systematic desensitization (counterconditioning to reduce anxiety by pairing negative events with complete physical relaxation—a positive event to extinguish the negative quality), or covert sensitization (the pairing of anxiety-producing stimuli with those that have pleasant associations); and (4) **operant conditioning,** whereby selected behaviors are immediately reinforced and systematic **schedules of reinforcement** have been consciously determined in advance. The goal of behavior therapy is to change behavior in general by increasing or decreasing specific behaviors, using empirically based treatments.

Implications of the Behavioral Approach for Helpers

The direct helping relationship is not considered important to the behavioral approach; the human relationship is not a variable in changing behavior unless, for some reason, the helper has tremendous reinforcement value to the helpee. The helper is seen more as a behavior engineer: objective, detached, and professional. In fact, the helper often serves merely as a consultant and does not have direct contact with the helpee, except perhaps for initial observation. There are even cassette tapes available for specific behavioral techniques. Where there *is* verbal contact, the helper is highly trained to use systematically concrete, verbal reinforcement. The helper is also highly trained in observational and evaluation skills. Because of the focus on observable, overt behavior, the behavioral helper is able to determine what interventions have been effective and when to terminate therapy by noting the presence or absence of specific target behaviors. Thus the helper is active and directive, often functioning as a teacher or consultant-trainer.

Because the behavioral approaches are specific and relatively open to measurement and evaluation, they have helped to move counseling and helping from the "art" end of the continuum toward the "science" end. Focusing on behavioral outcomes appears to mitigate dependency relationships and results in short-term treatment. As helpers we can focus on observable behavior (rather than talk about feelings and thoughts, which are difficult to measure); use techniques that enable us to evaluate the results of our interventions; specify goals, strategies, and outcomes of helping and thus be more accountable and avoid running the risk of imposing our subjective values and attitudes on helpees; focus on what the helpee is doing and can do as opposed to what the helpee cannot or should not do, thereby emphasizing the positive rather than negative aspects of individual behaviors; and deal only with observable behaviors and avoid risking questionable interpretations of subjective data.

☐ Cognitive-Behavioral Theory

Cognitive-behavioral approaches to helping deal with rationality, the thinking processes, and problem solving. They focus on the helpee's appraisals, attributions, belief systems, and expectancies and on the effects of those cognitive processes on emotions and behaviors. They are instructive, directive, and verbally oriented (as opposed to an approach like Gestalt, which is nonverbally oriented). Many of the vocational counseling approaches fall into this domain, with an emphasis on testing, synthesis of a variety of collected data, and rational decision making. The major philosophical assumption of cognitive-behavioral theory is that by changing people's thinking one can change their belief system, which in turn changes their behavior.

An emerging cognitive-behavioral theory known as **constructivism** focuses on the interdependence of thinking, feeling, and behavior, rather than on the supremacy of thinking. These theorists believe that people are active creators and construers of their realities and they focus particularly on how people process new information in order to adapt to environmental demands. This view is attempting to integrate emotional, cognitive, and biological development with informational and systems theories.

Three other major representatives of cognitive-behavioral theory are rational-emotive therapy (RET), reality therapy (RT), and cognitive-behavioral therapy.

Rational-Emotive Therapy

This approach to helping was developed by Albert Ellis in the mid-1950s and is based on his belief that people need to change their way of thinking (cognitive restructuring) to correct faulty (irrational) thinking.

Rational-Emotive Therapy's Major Principles of Helping

Ellis believes that people must take full responsibility for themselves and their own fates. He maintains that although people are influenced by biological and environmental factors, they are not controlled by them. Rather, their thought processes mediate between those factors and their emotions. People perceive, think, feel, and behave simultaneously. They can learn to control what they feel and do, to a great extent. Ellis further posits that people are born with a tendency to be both rational and irrational, with both self-sabotaging and self-actualizing capacities. They are greatly affected by social conditioning, so although there are biological, genetic determinants of rational and irrational thinking, people primarily teach themselves to become more and more irrational by incorporating the world's irrationality into their belief systems.

Ellis postulates an A-B-C theory: A = the activating experience that the client wrongly believes causes C = consequences; B = the client's belief system, which is the intervening variable really causing C. It is B, the client's belief system, that needs to be restructured. Ellis identified 12 irrational ideas (listed in Chapter 6) one keeps repeating to oneself that cause faulty thinking.

The following example demonstrates faulty thinking.

Because it would be highly preferable if I were outstandingly competent, I absolutely should and must be. It is awful when I am not, and I am therefore a worthless individual. Because it is highly desirable that others treat me considerately and fairly, they absolutely should and must and they are rotten people who deserve to be utterly damned when they do not. And because it is preferable that I experience pleasure rather than pain, the world absolutely should arrange this, and life is horrible and I can't bear it when the world doesn't.

Ellis claims that these are the kinds of things we tell ourselves that make us upset—but we can learn to give ourselves different messages.

The major constructs of this theory are the following:

1. Problems are caused by irrational beliefs that result in dysfunction.
2. People are capable of changing their belief systems by learning how to refute their irrational beliefs.
3. People are biologically and culturally predisposed to choose, to create, to relate to others, and to enjoy, but they also have inborn propensities to be self-destructive, evasive, selfish, and intolerant.
4. Emotional disturbance results from people's continual irrational thinking, their refusal to accept reality, their insistence on having things the way they think they should be, and their self-absorption.

Rational-emotive therapy is a teaching approach to helping people achieve personality change. By instructing, giving information, teaching **imagery techniques,** and assigning homework, the therapist helps change the helpee's irrational belief system. Thus, the technique is both cognitive (teaching) and behavioral (role playing, homework assignments). RET techniques are designed to change behavior by changing thinking, which in turn will help the client to feel better. Clients are taught to understand themselves, to understand others, to react differently, and to change their basic personality patterns and philosophy by correcting faulty thinking.

Implications of the Rational-Emotive Approach for Helpers

In this approach, the helping relationship is cognitive and directive in that helpers exhort, frustrate, and command helpees to get them to analyze their thoughts and learn to rationally restructure their belief systems. The helping relationship depends on the potency of helpers and on their ability to communicate that power to helpees, rather than on warm, empathic, reciprocal relationships. RET therapists accept clients as fallible humans without necessarily giving personal warmth. They criticize and point out behavioral deficiencies and encourage clients to use more self-discipline. They discourage dependency and often use impersonal, highly cognitive, active-directive homework assigning and discipline-oriented techniques such as didactic discussion, bibliotherapy, and audiovisual aids.

This approach has many implications for addressing the cognitive domain and helping to integrate it with the behavioral and affective domains. Helpers should try to recognize the interrelation of reason and emotion; to recognize clients' responsibility for declaring and reassessing their own belief systems; and to use a methodology that quickly exposes the relationship of feelings, thoughts, and behavior.

Reality Therapy

This branch of cognitive-behavioral theory, developed by psychiatrist William Glasser, is, like rational-emotive therapy, rational, logical, and learning-oriented. Reality therapy explores the client's values and behavioral choices, exposing inconsistencies and enforcing responsibility for those choices.

Reality Therapy's Major Principles of Helping

According to this theory, people have two basic psychological needs: the need to love and be loved and the need to feel worthwhile to themselves and to others. The theory is interpersonal in that one must be involved with others in order to meet those needs. The goal of reality therapy is to assist people to make responsible choices (involving consistency between value systems and behavior) and to meet the two basic psychological needs without depriving other people of the opportunity to fulfill their own needs. People are seen as capable of assuming responsibility for themselves and capable of rational thought and behavior. They have no excuse for not exercising these attributes.

Glasser's more recent work has stressed helping people take more effective control of their lives by helping them choose effective, responsible behaviors to fulfill the following needs: survival, belonging, power, fun, and freedom. These five needs are genetically determined, and our behavior is designed to control our environment in ways that will satisfy them. Thus, current reality therapy theory is based on control theory. There are two fundamental ways to control the world to meet our needs: (1) to perceive what in the world can possibly satisfy needs—input, and (2) to act upon or control what is perceived to satisfy needs—output.

The practice of reality therapy involves a candid, human relationship through which helpers teach helpees to accept responsibility for themselves by analyzing inconsistencies among their goals, values, and behaviors. Reality therapy is a direct approach; it deals with the present. It is based on eight steps:

1. Make friends and ask clients what they want.
2. Ask clients what they are choosing to do to get what they want.
3. Ask if their behavioral choice is working.
4. If it is not, as is almost always the case, help them to make better choices.
5. Get a commitment to follow the better choices that have been worked out in the previous planning phase.
6. Do not accept excuses for failure to carry out the plan; if the plan is impracticable, replan.
7. Do not punish, but ask clients to accept reasonable consequences for their behavior.
8. Do not give up.

Implications of the Reality Therapy Approach for Helpers

The reality therapist encourages, suggests alternatives, praises positive behavior, confronts inconsistencies openly and directly, and cares enough about the helpees to reject behaviors that prohibit them from meeting their basic psychological needs. Helpers are concerned about present behavior, what helpees are currently doing, and what the consequences of alternative choices might be. They refuse to engage in self-defeating conversations about negative experiences and symptoms.

The relationship between the reality therapist and the client is crucial; it must be a warm, honest, personal involvement. Through this relationship, the client learns to feel loved by the therapist, learns to love the therapist, and learns to feel worthwhile because the therapist focuses on what the client is doing, on the client's approach behaviors rather than on avoidance behaviors. (This process is similar to the psychoanalytic concept of transference, although it is not acknowledged as such.) This relationship differs from the client-centered relationship in one major aspect: the helping relationship in reality therapy is indeed judgmental in that the reality therapist judges the value of the client's behavior in terms of the therapist's perceptions of reality. For example, if the client persists in self-defeating behavior, the therapist may say, "That behavior is really crazy because it will keep you from getting what you say you really want."

The major implication of reality therapy for helpers is that honest, intense involvement of the helper and the helpee is required for success. Helpers must examine the inconsistencies between clients' actions and their values and needs, focus on action (what clients are doing rather than what they are thinking or feeling), and insist on clients' assuming responsibility for their choices and the consequences of those choices. Helpers facilitate helpees in meeting their needs and gaining control of their world.

Cognitive-Behavioral Therapy

As developed by Beck (1976), this therapy represents a major theoretical orientation today. Beck postulates three major components of a theory of emotional disorders:

1. Negative, automatic thoughts that disrupt one's mood and cause further thoughts to emerge in a downward thought-affect spiral
2. Distorted reality based on systematic logical errors, such as the following:
   a. Arbitrary inference: inferring a conclusion from missing, false, or irrelevant evidence
   b. Overgeneralization: concluding from one specific negative event that another is therefore more likely
   c. Selective abstraction: focusing on certain aspects of a situation and ignoring others
   d. Magnification or minimalization: thinking the worst of every situation or else refusing to acknowledge its importance
   e. Personalization: relating an external circumstance to oneself when there is no basis for that relationship
   f. Dichotomous thinking: all-or-nothing extremism
3. Depressive schemas in which one's assumptions about the world represent the way one organizes past experience and are the system by which incoming information about the world is classified

For Beck, problems arise from distortions of reality based on faulty premises and assumptions. These distorted appraisals lead to specific emotions and, therefore, one's emotional response is consistent with the distortion, not with reality.

The three major stages of cognitive-behavioral therapy are (1) eliciting thoughts, self-talk, and the client's interpretations of them; (2) gathering with the client evidence for or against the client's interpretations; and (3) setting up experiments (homework) to test the validity of the client's interpretations and to gather more data for discussion. This is an active therapy in which the therapist collaborates with the client in the here and now. It is a verbal therapy, and each session establishes an agenda, structures the therapy time, summarizes periodically what is happening, questions the client, assigns homework, and asks the client to sum up the session. Beck emphasizes the necessity of accurate empathy, warmth, and genuineness in the helping relationship; he says rapport, collaboration, and mutual understanding are important. Some specific techniques are cognitive rehearsal, questioning, searching for alternatives, monitoring thoughts, reality testing, thought substitution, and teaching coping skills and self-control techniques.

☐ Transactional Analysis Theory

Transactional analysis (TA), developed by Eric Berne, is an approach that falls midway between the phenomenological and the cognitive approaches. It focuses on reality, is logical and learning-oriented like the cognitive-behavioral approaches, and uses many of the same techniques as Gestalt therapy, emphasizing the client's perspective. However, it emphasizes cognitive goals more than Gestalt therapy does, in that it strives for intellectual as well as emotional awareness.

Transactional Analysis Theory's Major Principles of Helping

This system divides the personality into the parent, the adult, and the child **ego states**. They are different from the psychoanalytic ego states (superego, ego, id) in that the TA ego states are observable, conscious, and segregated from each other. The parent ego state consists of "tapes," developed from original interactions with parents and societal models, of all the rules and caring the person has ever experienced. The adult ego state is in touch with the outside world and with logic, is computing and information giving, and is responsible for solving problems and incorporating facts and data from without. The child ego state is in touch with all the feelings, with what's going on within the person.

Healthy individuals are able to function appropriately in each of these ego states and are conscious of which ego state they are functioning in. They are aware of their interpersonal relationship patterns and how they structure time (by withdrawal, rituals, procedures, pastimes, games, or intimacy), and they are encouraged to learn how to achieve genuine intimacy by recognizing their manipulative styles and becoming aware of their **scripts** (early childhood decisions that determine roles they will play in later life), **rubberbands** (current feelings causing overreactions linked to feelings from earlier events in one's life), and **sweatshirts** (messages that people advertise, usually without awareness, such as "kick me"). People are seen as capable of assuming self-responsibility and overcoming negative parental or cultural influences. People need "strokes" (responses from other people) in order to grow and live, and they are taught how to assume responsibility for both obtaining and giving necessary strokes.

The transactional analysts believe that there are four major lifestyle positions:

1. *I'm not O.K., you're O.K.:* The position into which we all are born and that results in the need to learn that we are O.K. This is a position of the child.
2. *I'm O.K., you're not O.K.:* The position that, from a child's viewpoint, is occupied by parents or adults with a strong need to control other people. This is the parent position.
3. *I'm not O.K., you're not O.K.:* The position of a person who has no confidence in himself or herself nor in anyone else. This is a position of the child.
4. *I'm O.K., you're O.K.:* The position of an individual who is able to feel confident about herself or himself and others and to feel capable without needing to control and manipulate others. This is the adult position.

TA is oriented to the present, although it acknowledges the past and helps plan for the future. The goals of TA are to match intentions with behavior and to integrate feelings, thoughts, and behavior by (1) examining and updating the parent, (2) freeing the child from its "not O.K." position, (3) teaching the adult to function without being "contaminated" by the parent or child, and (4) developing intellectual and emotional awareness, thereby improving transactions. Clients are taught the principles of TA instructionally, via diagrams, lectures, demonstrations, and readings. They learn to recognize their own and other people's games, **rackets,** parallel and crossed interactions, use of strokes (reinforcements), scripts, and counterscripts.

Thus, TA examines relationships by analyzing communication. Both content and process are emphasized and Gestalt techniques (for example, role playing and dialogues) are the major methodology in addition to the verbal analysis of transactions and relationships. TA is a contract therapy in that each transaction between the therapist and client aims toward establishing a contract (whereby the client agrees to perform a specific behavior that will benefit all three of the client's ego states and that will resolve a conflict or unfinished business) and obtaining commitment to that contract.

Implications of the Transactional Analysis Approach for Helpers

The TA helping relationship is directive in that the transactional helper is a leader who gives clues and hunches and shares fantasies with the client to activate the client's adult. There is no advising, no diagnosing, no "why" questioning. The helper can help the client by emanating **protection** (it's safe for you with me), **permission** (you can be you with me; you can feel and behave as you want), and potency (I am capable, have the power of curing you). Like Gestalt therapy, TA is typically a group approach to helping.

TA provides a comprehensive, teachable model that is appropriate for all age groups and that provides a nontechnical vocabulary and concepts for understanding interpersonal relationships. In addition, helpers focus on communication styles and messages and on effective ways to teach better communication behaviors; they integrate aspects of behavioral, reality, and Gestalt therapies; and they provide short-term, flexible helping that deals equally with the affective, cognitive, and behavioral domains.

☐ Integrative Theories

Most helpers use principles, concepts, and techniques from a variety of different theoretical viewpoints and, therefore, have an eclectic or integrated approach to helping. This approach requires some flexibility and versatility on the part of the helper, as it is not a pure approach based on a single theoretical construct. It requires the helper to appreciate that each theory has limitations as well as something of value regarding the understanding of human behavior. A helper who uses an eclectic approach must always strive to be consistent and comprehensive in integrating the different approaches and must be clear about how the approach is or is not related to the theory from which it is derived. Obviously, the helper must carefully select approaches that are appropriate for the client he or she is helping. Thus, an eclectic approach requires heightened self-awareness, so that helpers understand why certain viewpoints and approaches appeal to them and others do not. Furthermore, it is necessary for helpers to understand mind-body connections so as to consider the reciprocal influence of somatic and psychological conditions.

The eclectic approach, then, presupposes that helpers are continually open to and searching for new understanding of human behavior and for more effective techniques. This implies that they are actively involved in continual professional training. The current professional literature indicates a major shift toward integration and eclecticism in the practice of counseling and psychotherapy (Brabeck & Welfel, 1985; Ivey, 1991; Mahrer, 1989; Okun, 1990). Most authors caution against a "sloppy eclecticism" and urge synthesis of disparate views from an integration of theory based solidly on supportive empirical study.

Feminist Therapies

An approach rather than a unified theory, feminist therapies cut across the different available schools of theory. Feminist therapists focus on understanding gender as both a cause and a consequence of women's experiences in a male-dominated culture. They expose the limitations and constraints of the traditional normative value bases of existing psychological theories. Thus, their reason for being is to challenge and question the attitudes toward women of the prevalent psychological theories, which have been largely developed and practiced by male theorists and therapists. Feminist therapists believe that those theories advocate the maintenance of the status quo of a male-dominated, hierarchical society.

There are two major differences between feminist therapy and more conventional types. Feminist therapy is based on a relationship of equals between helper and helpee, who view women's problems as being inseparable from society's oppression of women. The second difference is an emphasis on social, political, and economic action as a major aspect of the helping process. Feminist therapists believe in stating their values at the beginning of the helping relationship and using those values deliberately in modeling and interpreting.

The kinds of helping goals commonly ascribed to feminist therapy include androgyny (the complementary coexistence of feminine and masculine characteristics within both males and females), equal-in-power relationships between men and women, acceptance of one's body image "as is," and career choice unbiased by sex (Dworkin, 1984). Feminist therapies range from cultural to radical feminism. The former focuses on gender differences and believes that each individual helpee needs to determine for herself or himself what sex-role options represent the best personal choice—that helpees should be exposed to feminist values but that the choice of values should be free and individual. Radical feminism minimizes gender differences and advocates equally active social analysis and change strategies, and individual development.

Feminist therapy cuts across the affective, cognitive, and behavioral domains. It is experiencing an ongoing development of theory; with the emergence of senior female behavioral science theorists such as Eichenbaum and Orbach (1983), Gilligan (1982), Hare-Mustin (1983), and Miller (1984), we are seeing more widespread attention devoted to the study of female development and types of helping particularly applicable to the female experience.

Cross-Cultural Models

Like feminist therapies, cross-cultural models do not subscribe to any one theoretical model but strongly question the relevance of mainstream theories to clients of different cultural backgrounds. Different elements of the clinical theories may apply to different cultural groups, and a culturally aware helper is appreciative of cultural diversity, the impact of nondominant socialization on minority cultures, acculturation issues, and the culture-specific variables of counseling theory and strategies, and is open to multiple perspectives (Okun, 1990; Pedersen, 1988; D. W. Sue, 1981; S. Sue, 1988).

Multimodal Therapy

As developed by Arnold Lazarus (1976, 1981, 1987, 1989), multimodal therapy is a comprehensive, systematic eclectic approach. Lazarus is perhaps the most articulate spokesman for an eclectic view. He describes a flexible, personalized approach to helping in which the helper uses a combination of techniques from different theoretical approaches without necessarily subscribing to their principal beliefs. The selection of techniques fits the treatment to the needs and individual characteristics of each client.

Theoretically speaking, the multimodal therapy approach is based on social learning principles, the reciprocal interaction between personal and environmental variables. Lazarus (1981) believes that personality is formed, maintained, and altered through many processes: classical and operant conditioning; modeling and vicarious learning; thoughts, feelings, images, and sensations; and unconscious processes such as interpersonal distortions and avoidances and metacommunications.

This approach postulates seven specific, interrelated aspects of human personality that need to be attended to by the helper: behavior (overt behaviors, observable and measurable); affect (emotions, moods, strong feelings); sensation (seeing, hearing, touching, tasting, smelling); imagery (created mental pictures); cognition (ideas, values, opinions, attitudes); interpersonal relationships (interactions with others); diet/drugs (exercise, nutrition, drugs). Hence the acronym BASIC ID. The helper covers as many of these modalities as possible, using techniques from whichever approaches focus on them. For example, phenomenological techniques may deal with the affect modality, whereas cognitive restructuring can deal with the cognition modality. The effectiveness and durability of therapeutic results are directly proportional to the number of modalities addressed, according to Lazarus.

Lazarus is really proposing a technical eclecticism rather than a theoretical eclecticism. He believes that helpers can successfully use techniques from different models without having to subscribe to the theories from which these techniques are derived. It is the effectiveness of the technique in treating a client's particular problem that is the concern of the therapist, not the theoretical viewpoint about the cause and meaning of the client's problem (Lazarus, 1987).

Developmental Counseling and Therapy

Developed by Allen Ivey, developmental counseling and therapy (DCT) organizes and integrates traditional developmental theories (Erikson's life-span theory, Haley's family life cycle theory, Bowlby's attachment theory) with more contemporary developmental theories (minority development theory and developmental counseling and therapy theory) into clinical practice and research. According to Ivey (1991), since the purpose of all helping interventions is to facilitate human development over the life span, the helper's task is to focus on the helpee's unique way of viewing the world. DCT enables the helpee to expand his or her cognitive and behavioral repertoires so as to develop new alternatives for life choices as well as adaptation skills. Ivey's unique contribution is his elaboration of intentional skill interviewing (specific questioning strategies) and developing the level of intervention in accordance with the helper's personal emotional development, level and style of cognitive development, and multicultural awareness and development. Thus, a systematic four-level treatment plan can range from sensorimotor work in nutrition and bodywork through concrete action, such as dieting and associated behavioral management, to formal examination of the self and then dialectic family systems work. Behavior modification and Gestalt strategies might be used at the sensorimotor level, rational emotive therapy and reality therapy might be used at the concrete-operational level, person-centered, existential or psychodynamic therapy might be used at the formal-operational level, and feminist therapy and family systems therapy at the dialectic-systemic level.

If we want to provide effective human services for a broad population, we must be able to respond to the variety of needs it presents. Adherence to a single theoretical view may meet the helper's needs more than the helpee's. To date, there has been no solid, replicable research to prove that any one approach is the best for all helpers and helpees.

☐ The Systems Perspective

No survey of helping theory is complete without reference to the systems perspective, which provides a larger context within which to help an individual. Its major premises are that (1) an individual is a system in himself or herself, comprised of interacting components or subsystems such as the cognitive, affective, and physiological arenas, and (2) an individual is a component of larger social systems such as the family, school, work place, and community. An individual's problems have meaning to the larger social system. All behavior is relational and communicative and there is a mutually reciprocal influence between one's problems, one's life circumstances, and the interactional patterns of family, school, work place, and community. So, for example, if a freshman in college is excessively homesick and becomes depressed enough to need to go home, the problem may turn out to be that the student feels responsible for ensuring that her parents' marriage survives, and perceives that if she goes home she can be the stabilizing "third leg" in a shaky two-person relationship. The focus of the helping process in this case would be more on the dysfunctional family (interpersonal relationships) than on the individual "sickness."

Thus, the identification and clarification of an individual's problems require consideration of both individual and system contributing factors. An individual's problems affect his or her family just as the family affects the individual. If, for example, an individual practices distorted thinking that affects his or her feelings, those feelings will affect interpersonal relationships with family or colleagues that, in turn, may cause problems that reinforce and intensify the distorted perceptions. This is the notion of **circular causality** (events are related through a series of interacting loops or repeating cycles and codetermine each other). It differs from the notion of linear causality (events are related through sequential development, and preceding events influence later events) underlying the major theories of helping.

Systems theory provides two important tenets for helpers to keep in mind: that a change in one component of a system (within the individual or the larger social system) will effect changes elsewhere in the system, and that the needs and goals of a larger system take precedence over those of a subsystem or component. How an individual functions and makes choices is greatly influenced by internal and external systems such as internal thinking, affective, and physiological systems and family, school, and other social-influence systems. Likewise, any growth or change on the part of the individual has an impact on the social systems within which the individual functions.

The systems perspective serves as a framework within which to consider individual development, functioning, and behavior change. See Okun (1984, 1990) and Okun and Rappaport (1980) for further discussion of this perspective.

☐ Summary

This chapter stresses the need to become aware of one's personal theory of helping and to differentiate between a day-to-day "theory of use" and a more conceptual "espoused theory." One's personal theory of helping will inevitably influence one's perception and assessment of the major formal theories of helping that we have presented in a brief overview: psychodynamic (classical, Adlerian, Jungian, and object relations), phenomenological (client- or person-centered and Gestalt), behavioral, cognitive-behavioral (rational-emotive, reality, and cognitive-behavioral therapies), transactional analysis, and interactive (feminist, cross-cultural, and multimodal therapy) developmental counseling and therapy theories, and the systems perspective. These theories provide the framework for understanding the strategies presented in the following chapter. Table 5.1 compares these major theories in schematic fashion; the following summary should also help to distinguish the broad outlines of each approach.

1. The psychodynamic approach emphasizes unconscious causes of behavior and early childhood experiences; it focuses on content more than on process.
2. The phenomenological approach emphasizes process more than content and stresses the helping relationship as a vehicle for change. This helping relationship provides a climate in which clients can explore their own feelings, thoughts, and behaviors to achieve both insight and behavioral change. This approach focuses on the present, as opposed to the past.
3. The behavioral approaches emphasize environmental consequences of behavior as determinants of behavior; they focus on the learning of new behaviors and the extinction of maladaptive behaviors. The process is one of identifying dysfunctional behaviors, planning new behaviors, and systematically arranging valued reinforcements for these new behaviors.
4. The cognitive-behavioral approaches are concerned with teaching new ways of thinking and with exploring and examining discrepancies between

Table 5.1 A comparison of major theories of helping

	Psychodynamic	Phenomenological Client-Centered	Phenomenological Gestalt	Behavioral	Cognitive-Behavioral	Transactional Analysis
Major principles	1. People have no free will	1. People have free will	1. People have free will	1. People have no free will	1. People have free will	1. People have free will
	2. Behavior is determined by biological and environmental factors	2. Locus of behavioral determinism is internal	2. Organism works as a whole within environment	2. Behavior is caused by environment and shaped by consequences	2. Behavior is determined by logical thinking and responsibility	2. People can choose their own behavior
	3. Neurosis stems from repressed infantile conflicts	3. Neurosis stems from incongruence between self-concept and environment	3. Neurosis stems from impasse between organism and environment	3. Neurosis results from maladaptive learning	3. Neurosis stems from irrational thinking and irresponsible choices	3. Neurosis stems from ego state contamination and exclusion
Therapy process	1. Directive	1. Nondirective	1. Directive	1. Directive	1. Directive	1. Nondirective
	2. Diagnosis and past experience important	2. Here-and-now experience important	2. Here-and-now verbal and body awareness important	2. Present behavior and reinforcement history important	2. Present behavior, thinking, and values important	2. Early script, present games, and transactions important
	3. Verbal	3. Verbal	3. Verbal and nonverbal; games	3. Verbal; activities	3. Verbal; activities	3. Verbal and nonverbal
	4. Transference in relationship important	4. Empathic relationship important	4. Honest, trusting, and supportive but confrontational relationship	4. Analytical, conditioning relationship	4. Instructional, involved relationship	4. Potent, protective, and permissive relationship

(continued)

131

Table 5.1 A comparison of major theories of helping (*continued*)

	Psychodynamic	Phenomenological Client-Centered	Phenomenological Gestalt	Behavioral	Cognitive-Behavioral	Transactional Analysis
Requisite therapist behaviors	1. Therapist neutral, benign, objective, but also empathic	1. Therapist honest, congruent, empathic, nonjudgmental, capable of unconditional positive regard	1. Therapist honest, open, confrontational but basically supportive	1. Therapist analytical, objective, observational, evaluative	1. Therapist involved and judgmental; confronts illogical thinking and irresponsibility	1. Therapist analyzes games, scripts, transactions, and ego states
	2. Interprets transference, resistance, and unconscious material	2. Needs ability to communicate the above qualities	2. Provides experience via verbal and nonverbal games	2. Analyzes goals, directs strategies, evaluates, provides positive reinforcement, arranges environmental contingencies	2. Rejects illogical thinking and irresponsible behavior	2. Provides experience via verbal and nonverbal games
	3. No contracts	3. No contracts	3. No contracts	3. Explicit contracts	3. Explicit contracts	3. Explicit contracts
Domain	Affective/cognitive	Affective	Affective	Behavioral	Cognitive/behavioral	Affective/cognitive/behavioral

132

values and behaviors. They also function in the present, rather than in the past.

5. Transactional analysis focuses on relationships (via analyses of communication) and lifestyles of clients, and aims at an integration of feelings, thoughts, and actions.

6. The integrative theories value the unique contributions of differing perspectives, and combine various helping approaches according to clients' situations and needs.

7. The systems perspective focuses on interpersonal relationships—that is, interactions. Its view of symptoms is that they both reflect and control interpersonal relationships.

Perhaps the reason that we have so many different approaches to helping is the great diversity that exists among people and their problems. No single theory answers all questions or satisfies all conditions. As helpers find themselves using approaches and viewpoints that are consistent with their personal theories and their personalities, they are bound to constantly refine existing approaches and create new ones. Feminist therapy is an increasingly significant approach to the helping relationship. It is critically important that we consider gender differences in our appraisal of helping theory and application, and in our support of research about the process and outcomes of the helping relationship.

One example of a nontraditional adaptation of conventional major approaches is the social-systems approach. This approach deals with the identification of systemwide problems rather than individual problems. Thus, the client is a system, such as a family, hospital, school, or correctional institution, rather than an individual within a system. The helper looks at the communication patterns as well as the structures (roles, rules, boundaries) of the system in order to understand the reciprocal influences of individuals and the contexts in which they function. Strategies derived from this approach include advocacy and change-agentry. Human relations skills are essential in applying these strategies because effective relationships between helpers and individuals within the system are crucial.

As the helping professions expand and mature, there is an increasing tendency toward a more open, multifaceted perspective of the major traditional theories and the currently emerging integrated theoretical views. This integration will allow helpers to consider all the aspects of human development, behavior, and change at different levels in different situations and ideally enable us to provide helping services to a greater variety of people in confusing, complex situations.

I hope that this overview of major helping theories will both heighten your interest in the theories and help you understand your own theoretical framework. The following reading list will help you further explore these views and approaches.

☐ References and Further Reading

General

Brammer, L., Shostrom, E., & Abrego, P. J. (1989). *Therapeutic psychology: Fundamentals of actualizing counseling and therapy* (5th ed.). Englewood Cliffs, NJ: Prentice-Hall.

Corey, G. (1990). *Theory and practice of counseling and psychotherapy* (4th ed.). Pacific Grove, CA: Brooks/Cole.

Okun, B. F. (1990). *Seeking connections in psychotherapy.* San Francisco: Jossey-Bass.

Patterson, C. H. (1986). *Theories of counseling and psychotherapy* (4th ed.). New York: Harper & Row.

Psychodynamic Theory

Psychoanalytic

Adler, A. (1927). *The practice and theory of individual psychology.* New York: Harcourt Brace Jovanovich.

Alexander, F. (1963). *Fundamentals of psychoanalysis.* New York: Norton.

Erikson, E. H. (1963). *Childhood and society* (2nd ed.). New York: Norton.

Erikson, E. H. (Ed.). (1968). *Identity: Youth and crisis.* New York: Norton.

Erikson, E. H. (1982). *The life cycle completed.* New York: Norton.

Freud, A. (1946). *The ego and the mechanisms of defense.* New York: International Universities Press.

Freud, S. (1943). *A general introduction to psychoanalysis.* Garden City, NY: Doubleday.

Freud, S. (1949). *An outline of psychoanalysis.* New York: Norton.

Hall, C. (1954). *A primer of Freudian psychology.* New York: Mentor.

Jung, C. (1928). *Contributions to analytic psychology.* New York: Harcourt Brace Jovanovich.

Jung, C. (1933). *Modern man in search of a soul.* New York: Harcourt Brace Jovanovich.

Jung, C. (1933). *Psychological types.* New York: Harcourt Brace Jovanovich.

Object Relations

Fairbairn, W.R.D. (1954). *An object relations theory of personality.* New York: Basic Books.

Guntrip, H. (1979). *Psychoanalytical theory, therapy, and the self.* New York: Guilford.

Kernberg, O. (1968). The therapy of patients with borderline personality organization. *International Journal of Psychoanalysis, 49,* 600–619.

Klein, M. (1932). *The psychoanalysis of children.* London: Hogarth Press.

Kohut, H. (1971). *The analysis of self.* New York: International Universities Press.

Mahler, M., Pine, F., & Bergman, A. (1975). *The psychological birth of the human infant.* New York: Basic Books.

Winnicott, D. W. (1965). *The family and individual development.* London: Tavistock.

Phenomenological Theory

Client-Centered

Carkhuff, R., & Berenson, B. (1967). *Beyond counseling and therapy.* New York: Holt, Rinehart & Winston.

Combs, A. (1989). *A theory of therapy: Guidelines for counseling practice.* Newbury Park, CA: Sage.

Combs, A., Avila, D., & Purkey, W. (1977). *Helping relationships: Basic concepts for the helping process* (2nd ed.). Boston: Allyn & Bacon.

Hart, J. (1970). *New directions in client-centered therapy.* Boston: Houghton Mifflin.

Rogers, C. (1951). *Client-centered therapy.* Boston: Houghton Mifflin.

Rogers, C. (1961). *On becoming a person.* Boston: Houghton Mifflin.

Rogers, C. (1980). *A way of being.* Boston: Houghton Mifflin.

Rogers, C., & Dymond, R. (1954). *Psychotherapy and personality change.* Chicago: University of Chicago Press.

Existential

Bugental, J.F.T. (1987). *The art of the psychotherapist.* New York: Norton.

May, R. (1961). *Existential psychology.* New York: Random House.

May, R. (1983). *The discovery of being: Writings in existential psychology.* New York: Norton.

Yalom, I. (1980). *Existential psychotherapy.* New York: Basic Books.

Gestalt

Fagan, J., & Shepherd, I. (Eds). (1970). *Gestalt therapy now.* Palo Alto, CA: Science and Behavior Books.

Koffka, K. (1935). *Principles of Gestalt psychology.* New York: Harcourt, Brace.

Perls, F. (1969). *Gestalt therapy verbatim.* Lafayette, CA: Real People Press.

Perls, F., Hefferline, R., & Goodman, P. (1951). *Gestalt therapy.* New York: Dell.

Yontef, G. (1982). Gestalt therapy: Its inheritance from Gestalt psychology. *Gestalt Theory, 4,* 23–39.

Behavioral Theory

Bandura, A. (1969). *Principles of behavior modification.* New York: Holt, Rinehart & Winston.

Bandura, A. (1977). *Social learning theory.* Englewood Cliffs, NJ: Prentice-Hall.

Franks, C., Wilson, G., Kendall, P., & Brownell, K. (1982). *Annual review of behavior therapy: Theory and practice,* vol. 8. New York: Guilford Press.

Kazdin, A. E. (1989). *Behavior modification in applied settings* (4th ed.). Pacific Grove, CA: Brooks/Cole.

Krasner, L., & Ullman, L. (Eds). (1965). *Research in behavior modification.* New York: Holt, Rinehart & Winston.

Krumboltz, J. (Ed.). (1966). *Revolution in counseling: Implication of behavioral sciences.* Boston: Houghton Mifflin.

O'Leary, K. D., & Wilson, G. T. (1987). *Behavior therapy: Application and outcome* (2nd ed.). Englewood Cliffs, NJ: Prentice-Hall.

Wolpe, J. (1958). *Psychotherapy by reciprocal inhibition.* Palo Alto, CA: Stanford University Press.

Wolpe, J., Salter, A., & Reyna, L. (1964). *The conditioning therapies: The challenge of psychotherapy.* New York: Holt, Rinehart & Winston.

Cognitive-Behavioral Theory

Beck, A. T. (1976). *Cognitive therapy and the emotional disorders.* New York: International Universities Press.

Beck, A., Rush, J., Shaw, B., & Emery, G. (1979). *Cognitive therapy of depression.* New York: Guilford Press.

Dobson, K. S. (Ed.). (1988). *Handbook of cognitive-behavioral therapies.* New York: Guilford Press.

Ellis, A. (1962). *Reason and emotion in psychotherapy.* New York: Lyle Stuart.

Ellis, A. (1973). The no cop-out therapy. *Psychology Today, 7,* 56–62.

Ellis, A., & Harper, R. (1975). *A new guide to rational living.* Englewood Cliffs, NJ: Prentice-Hall.

Ellis, A., & Whiteley, J. (Eds.). (1979). *Theoretical and empirical foundations of rational-emotive therapy.* Pacific Grove, CA: Brooks/Cole.

Glasser, N. (Ed.). (1989). *Control theory in the practice of reality therapy: Case studies.* New York: Harper & Row.

Glasser, W. (1965). *Reality therapy.* New York: Harper & Row.

Glasser, W. (1981). *State of the mind.* New York: Harper & Row.

Meichenbaum, D. H. (1977). *Cognitive behavior modification.* New York: Plenum Press.

Williamson, E. (1965). *Vocational counseling: Some historical, philosophical, and theoretical perspectives.* New York: McGraw-Hill.

Transactional Analysis Theory

Berne, E. (1961). *Transactional analysis in psychotherapy.* New York: Grove Press.

Berne, E. (1964). *Games people play.* New York: Grove Press.

Berne, E. (1972). *What do you say after you say hello?* New York: Bantam Books.

Goulding, R. I., & Goulding, M. (1978). *The power is in the patient: A TA/Gestalt approach to psychological thinking.* San Francisco: TA Press.

Harris, T. (1967). *I'm O.K., you're O.K.* New York: Harper & Row.

Integrative Theories

Brabeck, M., & Welfel, E. (1985). Counseling theory: Understanding the trend toward eclecticism from a developmental perspective. *Journal of Counseling and Development, 63,* 343–348.

Ivey, A. E. (1991). *Developmental strategies for helpers: Individual, family, and network interventions.* Pacific Grove, CA: Brook/Cole.

Mahrer, A. R. (1989). *The integration of psychotherapies.* New York: Human Sciences Press.

Okun, B. F. (1990). *Seeking connections in psychotherapy.* San Francisco: Jossey-Bass.

Feminist Therapy

Ballou, M., & Brown, L. (Eds.). (1991). *Personality theory and psychopathology: Feminist reappraisals.* New York: Guilford Press.

Cook, E. P. (1985). Androgyny: A goal for counseling? *Journal of Counseling and Development, 63,* 567–572.

Dworkin, S. (1984). Traditionally defined client, meet feminist therapist: Feminist therapy as attitude change. *Personnel and Guidance Journal, 62,* 301–306.

Eichenbaum, L., & Orbach, S. (1983). *Understanding women: A feminist psychoanalytic approach.* New York: Basic Books.

Gilligan, C. (1982). *In a different voice.* Cambridge, MA: Harvard University Press.

Hare-Mustin, R. (1983). An appraisal of the relationship between women and psychotherapy. *American Psychologist, 38,* 593–601.

Hare-Mustin, R. T., & Maracek, J. (1990). *Making a difference: Psychology and the construction of gender.* New Haven: Yale University Press.

Miller, J. B. (1984). Toward a development of theory of self. Paper presented at Wellesley College.

Multimodal Therapy

Lazarus, A. (1971). *Behavior therapy and beyond.* New York: McGraw-Hill.

Lazarus, A. (1976). *Multimodal behavior therapy.* New York: Springer-Verlag.

Lazarus, A. (1981). *The practice of multimodal therapy.* New York: McGraw-Hill.

Lazarus, A. A. (1987). The need for technical eclecticism: Science, breadth, depth, and specificity. In. J. K. Zeig (Ed.), *The evolution of psychotherapy* (pp. 164–178). New York: Brunner/Mazel.

Lazarus, A. A. (1989). *The practice of multimodal therapy.* Baltimore, MD: Johns Hopkins University Press.

Cross-Cultural Models

Okun, B. F. (1990). *Seeking connections in psychotherapy.* San Francisco: Jossey-Bass.

Pedersen, P. A. (1988). *A handbook for developing multicultural awareness.* Alexandria, VA: American Association for Counseling and Development.

Sue, D. W. (1981). *Counseling the culturally different: Theory and practice.* New York: Wiley.

Sue, S. (1988). Psychotherapeutic services to ethnic minorities: Two decades of research findings. *American Psychologist, 43,* 301–308.

Systems Perspective

Okun, B. (1984). *Working with adults: Individual, family, and career development.* Pacific Grove, CA: Brooks/Cole.

Okun, B. F. (1990). *Seeking connections in psychotherapy.* San Francisco: Jossey-Bass.

Okun, B., & Rappaport, L. (1980). *Working with families: An introduction to family therapy.* Pacific Grove, CA: Brooks/Cole.

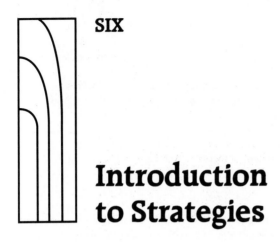

SIX

Introduction to Strategies

During the transition between stage 1 (development of the relationship) and stage 2 (strategy planning, implementation, and evaluation) of the helping process, the helper and helpee explore the goals and objectives of the helping relationship; they then focus on specific helping requirements and finally agree on their goals. Before this "substage" can be successfully concluded, helper and helpee must define the problem(s) to be solved and the nature of the help to be generated.

After the problem has been defined, it is possible to choose the appropriate strategy or combination of strategies to use in solving it. Variables of time, timing, setting, and the nature of the presenting problem will affect both how the transitional period is handled and which strategies are chosen to resolve the problem.

It is important to remember that a client's present problem needs to be addressed whether or not the resolution of more complex, underlying problems becomes a goal of the helping relationship. For example, if an employee is referred to a counselor for chronic tardiness, that issue needs to be resolved behaviorally immediately (so the employee won't be fired—unless he or she wishes to be); then the counseling process can focus on the employee's underlying or higher-order issues that may be being expressed via chronic tardiness.

☐ Strategies and the Three Main Problem Areas

Helping strategies can be categorized according to whether they deal with the affective, cognitive, or behavioral domains. Table 6.1 shows how the various strategies of the major helping theories address those areas. There have been several attempts (Bruce, 1984) in recent years to classify the strategies and philosophies of the various helping theories into a useful scheme—attempts that have been largely atheoretical and whose goals have been to bridge the gaps among

Table 6.1 Helping strategies as they relate to the three problem areas

Problem Area	Affective (emotion/feeling)	Cognitive (understanding/thinking)	Behavioral (action/doing)
Helping strategy	Client-centered therapy Gestalt therapy Psychoanalysis (Freudian, Jungian, Adlerian, object relations)	Reality therapy Rational-emotive therapy Multimodal therapy Transactional analysis	Behavioral therapy
Representative techniques	Gestalt experiments Responsive listening Imagery Sensory awareness Free association Dream analysis Interpretation	Transactional analysis Decision making Cognitive restructuring Reality therapy contracting Cognitive analysis and validation	Assertiveness training Reinforcement Contracts Modeling Systematic desensitization

(continued)

Table 6.1 Helping strategies as they relate to the three problem areas (*continued*)

Problem Area	Affective (emotion/feeling)	Cognitive (understanding/thinking)	Behavioral (action/doing)
Theoretical framework	Phenomenological	Cognitive-behavioral	Behavioral
	←---------------------- Eclectic ----------------------→		
	←-------- Psychodynamic -------- Transactional analysis --------→		
	←---------------------- Interpersonal adjustment ----------------------→		
	←---------------------- Developmental conflicts ----------------------→		
Kinds of problems	Anxiety Personal adjustment		
	Self-esteem	Problem solving Decision making	Behavioral problems
		Coping/mastery	

140

helping process efficacy, the helpee's immediate and longer-range needs, and the helper's needs and orientation.

Affective problems are those dealing with emotion, with self-awareness and awareness of others' feelings (for example, feelings of inadequacy or inferiority, or not being in touch with what and how you and others are feeling). For these kinds of problems, experiential strategies focusing on imagery, sensory awareness, and verbal and nonverbal expressions of feeling are effective.

Cognitive problems involve thinking (for example, decision making and problem solving). People who always seem to make the wrong decisions or who are afraid to make decisions or who refuse to accept responsibility for their actions can use help in this area. The didactic (instructional) strategies, which focus on step-by-step verbal processing of decision making, analyzing, and problem solving, are effective for such people.

Behavioral problems involve actions (for example, stopping smoking or some other habit, learning to be more assertive, or changing self-defeating behavior into behavior that elicits increased rewards). Behavioral strategies involve verbal and action-oriented instructions that arrange for environmental rewards and elicit behavior change.

Affective-cognitive-behavioral problems emerge in a broad variety of symptoms: they may be evidenced by depression, eating disorders, uncontrollable temper, or specific behavioral disorders. Cognitive-behavioral strategies include cognitive restructuring, assessment and instruction of client appraisals of self and others and of events, and training in coping skills or new, subvocal behavior.

These classes of problems and strategies are not always discrete—they may overlap or coincide. The nature of the problem(s), the nature of the helping relationship and situation, and the competence and skill of the helper all influence the choice of strategies.

When speaking of strategies in the helping relationship context, we are talking about overall approaches to achieving general or long-term goals. Strategies reflect the concepts and premises of specific theories or models for certain classes of problems. The techniques are specific applications of the strategies envisioned as the foundation of the respective problem-solving theories and models. Certain strategies and corresponding techniques work best in a particular type of situation. And although certain techniques may represent only one strategy, other techniques may be used by several different strategies.

There are times when the problem falls clearly into either the cognitive, affective, or behavioral domain and the range of strategies that can be used is apparent. For example, if the presenting problem is chronic tardiness, a behavioral problem, the strategy may be behavioral and the technique may be contracting, with or without cognitive restructuring. Often, however, the presenting problem is in a different domain from the underlying problem.

For example, a client was once referred to me by another counselor for the specific behavioral counseling technique called systematic desensitization. This client was unable to swallow solid foods; medical examinations were unable to find any organic cause for this condition. After several sessions of establishing a relationship and attempting to meet the client's expectations for systematic desensitization (which will be described later in this chapter), it became apparent that her suffering was rooted more in the affective domain than in the behavioral domain. She was unable to express any anger toward anyone, and she had accumulated a great deal of anger because of a recent broken engagement. She was losing weight rapidly and was endangering her health because she was eating only liquefied foods. Many sessions of phenomenological client-centered and Gestalt strategies were necessary for this client to be able to acknowledge her feelings and begin to express them appropriately. Eventually, she began to be able to swallow solid foods and to improve her interpersonal relationships. This is an example of the underlying problem (expression of anger) requiring a strategy quite different from the presenting problem (inability to swallow).

The following sections will provide you with a brief overview of strategies and techniques in each of the three major problem areas (affective, cognitive, and behavioral) and in the areas where they overlap (affective-cognitive and cognitive-behavioral). Please note that some of these strategies are more appropriate for experienced helpers than for beginning professionals and paraprofessionals. This introduction to the strategies will give you some idea of where your interests and inclinations lie and perhaps will suggest some directions for further study. Each section includes examples and exercises, and a reading list appears at the end of the chapter. As you work through the exercises, see which strategies are most comfortable and seem to make more sense to you.

☐ Affective Strategies

The theoretical rationales for affective strategies come from Carl Rogers's client-centered therapy and the Gestalt theory underlying Gestalt therapy. The focus is on self-awareness and experiencing feelings.

Rogerian client-centered therapy has contributed the basis for responsive listening communication skills. The helper, by communicating empathy, honesty, congruence, genuineness, and acceptance of the helpee, is able to create a nonthreatening climate in which helpees can explore their own feelings, thoughts, and behavior and gain some understanding of themselves and their world. Remember, Rogerian theory insists upon these environmental variables for the helpee to develop a positive self-concept. For this technique to be effective, the helpee must be able to perceive these feelings and attitudes of the helper. The technique of responsive listening may suffice as the only strategy needed in a helping relationship.

Gestalt strategies, on the other hand, specifically focus on awareness. Many helpers with theoretical orientations other than Gestalt nevertheless use Gestalt strategies to help clients achieve awareness. The purpose of Gestalt strategies is to reintegrate attention and awareness so that helpees can take responsibility for the *what* and *how* of their present behaviors.

Some of the rules for conducting Gestalt therapy are as follows:

1. Use the phrase "here and now" to focus on the present and on people who are here.
2. Use direct language, such as "I" instead of "it" and "I won't" instead of "I can't."
3. Allow no gossiping. The client should not talk about a person who is not present but instead should talk directly to the absent person by role playing.
4. Insist that clients own their own feelings, thoughts, and actions by using "my" and "I" and "I take responsibility for . . . ."
5. Direct the client to take action instead of imagining and thinking.

Games such as "Dialogue," "I take responsibility," and "Reversals" encourage present-oriented and responsibility-oriented verbal styles. In these games clients deal with an absent person by role playing, take both roles in a dialogue, play all the roles, including inanimate objects from a dream, and so forth.

The verbal techniques of Gestalt strategies are aimed at keeping the helpee in constant contact with what is going on. The following are the rules of verbal techniques:

1. Keep communication between helper and helpee in the "now" through the use of the present tense, and emphasize what is happening now. Use questions such as "What are you feeling now?" and "Are you aware that . . . ?"
2. Use the words "I" and "thou" to personalize and direct communication toward, not at, the listener.
3. Use the word "I" so that the helpee assumes more responsibility for his or her own behavior by substituting "I" for "it" (for example, instead of "The noise in the dormitory kept me from doing my homework" substitute "I did not do my homework").
4. Pursue the *what* and *how* instead of the *why* (for example, "What are you aware of now?" and "What are you experiencing now?" instead of "Why do you feel . . . ?"), which helps lead the client away from endless explanations, speculations, and interpretations.
5. Don't gossip. This rule promotes the expression of feelings and encourages the helpee to deal directly with people. If the people he or she is discussing are not present, the client is encouraged to talk directly with them using the **empty seat** or some other device.
6. Change questions into statements, which helps prevent manipulative games and encourages the helpee to take responsibility for and deal directly with issues.

The preceding rules are based on the following guidelines (Levitsky & Perls, 1970):

1. Live now; be concerned with the present rather than with the past or future. We spend too much time daydreaming about the past or future, and this habit distracts and detracts our energies and awareness from the present.
2. Live here and deal with what is present rather than what is absent. One of many avoidance strategies we use is to focus on what is missing rather than on what we have, on who is missing rather than who is here.
3. Stop imagining and experience the real. Imagining takes us away from what *is* and blocks our experiencing and awareness. We sometimes lose sight of what is real for us.
4. Stop unnecessary thinking, and taste, see, and feel. When was the last time, for example, that you ate an orange without thinking about the concept of orange but just sensing every feeling, taste, and smell? We have allowed thinking to block out our senses, and we need to take the time to get back in touch with our senses.
5. Express rather than manipulate, explain, justify, or judge. Learn to express yourself directly, to ask for what you want, to accept yourself and others for what they are, not for their verbal competencies.
6. Expand your awareness by giving in to unpleasantness and pain as well as pleasure. True awareness includes negative experiences as well as pleasurable ones, and, if we use our energies to block out the negative, we will also lose some of our ability to sense the positive.
7. Accept no "should" or "ought" other than your own, and follow no idol. The words "should" and "ought" have caused more difficulties than just about any other words in the English language. We must assume responsibility for our rules and mores and our own behaviors.
8. Take full responsibility for your actions, feelings, and thoughts. This is the essence of maturity in Gestalt thought. We must stop blaming others and situations and take full advantage of our autonomy and the choices that do exist within any situation.
9. Surrender to being you as you are. Accept yourself for who you are and what you are and not what you or others think you should be.

The following excerpt is an example of the application of a Gestalt strategy.

Client: I'm upset with my fiancé because he decided where we would live next year without talking it over with me. We're going to Des Moines, and that's over a thousand miles away!

Helper: You feel angry because he is making important decisions without consulting you.

Client: Well, yes. I don't want to go so far away. My mother lives all alone here, and we're very close. If I go so far away, I won't be able to see her and she needs me.

Helper: Are you aware that your right hand is tightly clenched?

Client: Oh . . . yes, I guess I'm more upset than I realized.

Helper: Let's try something to see if we can find out what this is all about. How about letting your right hand be the Pam who doesn't want to go away from your mother and your left hand be the Pam who wants to go with your fiancé. See if you can have your two hands talk to each other.

Client: Well, I'll try. *(shaking right fist)* Listen, you, you know you're afraid to leave your mother . . . she won't be able to manage without you and you don't really know if you can get along without her.

Helper: Now be the other Pam.

Client: Come on, now. I'm a big girl, and I certainly can make it on my own. Besides, I love Ron and I want to marry him, and that means I go where he goes.

Helper: See if you can talk directly to your other hand. Tell her what you feel.

Client: *(still the Pam who wants to go)* I'm annoyed at you for always getting in my way. You're a scaredy-cat and you always foul things up by getting angry when you don't want to do something. *(now the right-hand Pam, who doesn't want to go)* I don't want to go. I've never been that far away before.

Helper: What are you feeling now, Pam?

Client: I'm feeling scared, but also a little bit excited about the possibilities of a new life.

This short dialogue experiment helped Pam become aware of her true feelings and the real issues she is struggling with. Although there is reason for Pam to deal with and work through her feelings about her fiancé, it became clear that the underlying problems dealt with Pam's dependence/independence relationship with her mother.

The technique demonstrated in the preceding example is called the *dialogue game*. It can take place between two "aspects" of the helpee or between the helpee and another person with whom he or she is experiencing some kind of continued conflict. Other Gestalt techniques involve the use of imagery and sensory awareness, focusing on the relationship between verbal and nonverbal behavior ("You say you are angry, yet you are smiling"), acting out fantasies (playing all the roles of animate and inanimate parts of a fantasy), repeating and exaggerating verbal or nonverbal behaviors ("Can you stay with that feeling? Exaggerate your leg swing and repeat what you just said in a louder voice . . . louder . . . louder"), playing out projected roles by doing to others what one does to oneself, and completing unfinished business through active role playing. Gestalt therapists also ask helpees to play out dreams in the same way that they might play out fantasies. Any part or piece of the dream or fantasy is considered an aspect of the helpee, a metaphor to understanding what is happening in the here and now.

Important Gestalt questions include "What are you experiencing now?" "Where are you now?" "What do you want to do?" "What are you doing now?" and "What are you avoiding?" Helpees are encouraged to use "I" messages by completing such statements as "I am aware that . . . ." "Now I feel that . . . ," and "I notice that. . . ." Examples of directives that Gestalt helpers use include "Say *I* instead of *it*"; "Get a sense of your strong part"; "Be more specific"; "Say this again . . . again . . . now exaggerate it"; "Tell your strong part what it should do"; "Act stupid"; and "Act as if you don't care." Frequently, the Gestalt therapist will share his or her experiencing of what the client is doing at the moment; for example, "I'm aware that you're jiggling your foot while you talk about this" or "My hunch is that you're feeling scared, as if you want to run away and hide." The goals of the affective strategies are to develop feelings and self-awareness, using techniques of responsive listening and Gestalt experiments. The focus is on the present and the here and now.

When to Use Affective Strategies

Strategies that use responsive listening and focus on the development of a genuine, empathic relationship are appropriate for individuals who are unable to express their feelings and are unable to have close, meaningful personal relationships with family or friends. The technique of responsive listening may suffice as the sole strategy in a helping relationship whose goal is the development of self-concept in the helpee. If a helpee who has difficulties with interpersonal relationships is able to develop an honest, close, meaningful relationship with the helper, the experience is irreversible. And once having attained this type of relationship with one person, the helpee will be better able to achieve a close relationship with another. The responsive listening approach is also appropriate for informal and short-term relationships, when having someone listen to and understand one's concerns is helpful in and of itself.

Gestalt techniques are particularly effective for people who lack awareness of the *how* and *what* of their present behavior, people who refuse to take responsibility for themselves and their lives, people who interact rigidly and in a ritualized manner with their environment, people who dwell on past unfinished business or on future rehearsing, and people who seem split in two because they deny or exclude part of themselves. Gestalt techniques are also effective with children, who are more in touch with their fantasies and imagination than are older people. These techniques are generally not effective with people who do not want to develop awareness of their feelings, people who need information in order to make immediate decisions cognitively, people who have experienced a sudden crisis, and those who are unable to imagine and fantasize sufficiently to participate in the games and experiments.

To use Gestalt techniques effectively it is important for you to have actually participated in Gestalt games and experiments and thereby developed some confidence in your own capacities for self-awareness and taking responsibility. Try some of the following exercises to see how you feel about Gestalt techniques. You may wonder if people you are helping would think you were crazy for using these techniques, but you will find that if you have an effective, trusting helping relationship, helpees are usually willing to engage in novel techniques. (Actually, some of the communication skills exercises in Chapter 4 are Gestalt exercises, so you already have tried a few.)

Exercise 6.1   An effective beginning Gestalt exercise involves a group sitting in a circle; each person begins a three-minute monologue with "Now I am aware that . . . ." Try to get in touch with as much of yourself in the here and now as you can. Personalize pronouns and begin each sentence with "I" in order to focus on your self-awareness. For example, you might say, "Now I am aware that I am sitting in a hard chair with a pillow at my back. My legs are crossed, and I am typing this manuscript. My fingers are deftly moving over the keyboard, my eyes are on the draft, my shoulders are sort of slumped . . . " and so forth. (This exercise can also be performed in pairs.)

Exercise 6.2   Fantasizing is very much a Gestalt technique. In this exercise, each person in a small group is asked to think of a place where he or she feels especially comfortable, to visualize all the details of those surroundings, to get in touch with the sights, smells, and noises as well as the thoughts and feelings associated with this special place. Then, each person takes turns relating his or her scene in the present tense, personalizing pronouns. Other people in the group may ask "what" and "how" questions and may also ask the person sharing the fantasy to act out different aspects of it, playing out the animate and inanimate roles. For example, if I describe a scene at the beach, I may be asked to be the water, to be the sand, to be the sun.

Exercise 6.3   "Shuttling" is an extension of Exercise 6.2. Spend some time in your special place and then "shuttle" to the here and now and get in touch with the details of where you are right now, the sights, smells, noises, and other people. Then return to your special place. Shuttle back and forth between reality and fantasy and spend several minutes in each place. How hard is this shuttling? What do you feel? Where do you want to be? What have you become aware of?

Exercise 6.4   "Dialoguing" is a useful Gestalt technique. Imagine that one of your parents is sitting facing you. Describe this scene as precisely as you can, relating the emerging feelings as you face this parent. Then begin to talk aloud to this parent, using the first and second persons ("I" and "you"). Say whatever comes to your mind. Stay with the feelings that are emerging, and then switch seats when you choose and be your parent talking back to you. Go back and forth between your seat and your parent's seat until you feel as if you are ready to stop. If possible, share your feelings with your group and see if you have learned something about your feelings and relationship with this parent. If you are having a difficult time getting started, begin in your parent role, introducing you. Tell who you are and how you feel about your offspring, you.

Exercise 6.5   Try a variation of the preceding exercise by playing two different aspects of yourself. For example, you might be "strong" in one seat and "weak" in the other. Choose the aspects that are most comfortable for you and verbally share your thoughts, feelings, and experiences in one seat. When you finish, switch to the other seat and share your conflicting thoughts, feelings, and experiences from that perspective. Continue verbalizing and acting out this conflict until you feel you have some understanding (and perhaps experience some integration) of your different parts. This technique is termed "empty chair" by Gestalt therapists. It is a stimulating method of facilitating self-knowledge.

☐ Affective-Cognitive Strategies

The theoretical basis for affective-cognitive strategies is developmental psychodynamic theory. The primary goal is to bring unconscious material into the conscious realm so as to strengthen the ego in order for behavior to be based more on conscious thinking than on unconscious instincts. The personality becomes restructured by the achievement of insight (emotional awareness and cognitive understanding). The objective is to eliminate the crippling effects of internal anxiety, which both causes and results from repression, so that the client can live more fully in the present with inner peace and self-understanding. This enables clients to achieve more productive relationships and work functioning.

Techniques

The major techniques of psychodynamic helpers are free association, dream analysis, and interpretation. These are verbal techniques that allow helpees to proceed at their own pace to develop a transference relationship with the helper and work through unconscious conflicts. The purpose of free association and dream analysis is to allow helpees to become gradually aware of deeply unconscious material. The focus is on childhood experiences, to enable clients to understand the connections between the past and their current functioning.

Recent attention to the transference phenomenon focuses on the transfer of feelings, attitudes, and conflicts experienced in the past to current situations and relationships (Cashdan, 1988; Okun, 1990; Watkins, 1983). By recognizing the possibility of this phenomenon, helpers can become more aware of its impact on the helping process. Watkins posits five major transference patterns commonly found in the counseling relationship. The counselor may be perceived and treated as (1) ideal, (2) seer, (3) nurturer, (4) frustrater, or (5) nonentity. The perception of each type will affect client attitude and behaviors as well as the counselor's experience of the relationship. For example, if the client experiences and treats the counselor as a seer, he or she will expect expert advice and solutions, and the counselor may experience feelings either of omnipotence or of incompetence for being unable to come up with the answers. In such a case, the counselor needs to use strategies that address the client's past dependency needs and past relationships with "authority figures," and to focus on helping the client gain self-esteem and independence.

When to Use Affective-Cognitive Strategies

Psychodynamic techniques are in order for people who have persistent, deep-seated problems requiring restructuring of the personality. Unless you pursue psychoanalytic training, it is unlikely that you will have the opportunity to do more than provide support to this type of helpee, perhaps as a psychiatric aide.

The following exercises will help you experience free association and what it can mean to and for you.

Exercise 6.6   In pairs or small groups, take ten minutes for each person to look at a list of words (such as *red, blue, black, pink, white*) and say whatever comes to mind as a result of the stimulus word. After each person has taken ten minutes for this task, he or she can share reactions and observations. Other members of the group can then ask this person some questions and share their reactions.

Exercise 6.7   With a partner, recall every verbal slip you've made in the past two weeks. Then you and your partner should brainstorm all the possible unconscious meanings of those slips. See if you can help each other get in touch with what the slips symbolized.

Exercise 6.8   Recall a recent dream. Allow yourself to focus on one particular aspect of or character in the dream and attempt some free association. See what you can learn about the possible meaning of your dream.

Exercise 6.9   Look around the class and allow your gaze to rest on the person for whom you have the strongest dislike. Does this person remind you of someone else? Allow yourself to free-associate and see if you can achieve some insight about this transference.

Exercise 6.10   Look back over the significant authority figures in your life, such as parents, teachers, doctors, and employers. Think of one whom you idealized and then think of another you experienced as frustrating or rejecting. Can you separate the feelings and behaviors that were appropriate from those that were overreactions and projections? Do these people remind you of anyone else? Are there patterns in the way you perceive, feel about, and behave toward authority figures? How do current relationships mirror your past experiences? What kinds of feelings does this exercise elicit?

☐ Cognitive Strategies

Cognitive strategies emphasize rationality, thinking processes, and understanding. The theoretical foundations are information and decision-making systems.

Techniques

Decision-making techniques are used in the cognitive problem area because decisions are cognitive processes. It is important to help people learn decision-making skills so that they will have more freedom and control over their lives. We make decisions from the moment we get up in the morning to the moment we retire.

Although there are different models for decision making, the basic process recommended for helping relationships consists of the following steps:

1. *State the problem clearly.* Be sure that you have identified the problem that is causing difficulty. For problem solving to be effective, the problem must be accurately identified.
2. *Identify and accept ownership of the problem.* Unless the decision maker believes that he or she has a problem and has some power to effect a decision, the decision-making process is futile. People do not invest energy in decision making unless they have a stake in the outcome.
3. *Propose every possible alternative to the problem* (**brainstorming**). Often our options are limited. Brainstorming enables us to consider all possible options without judging them. It gives us more to choose from.
4. *Evaluate each proposed alternative in terms of implementation realities and hypothesized consequences (value clarification).* Here we have the opportunity to evaluate each of the alternatives proposed in step 3. Some we will automatically discard, either because they are impractical or because they violate our value system. Before discarding any item, however, we try to hypothesize its consequences.
5. *Reassess the final list of alternatives, their consequences, and the risks involved.* We review our final list, and for each alternative we review the steps involved and the likely consequences. We may eliminate additional alternatives in this step.
6. *Decide to implement one alternative.* Based on our previous assessments, we choose one alternative. We may even list some back-up alternatives.
7. *Determine how and when to implement the plan.* Here we spell out exactly what is needed to implement this decision, who needs to do what, when, where, what materials are required, and so forth. Decisions are often not carried out because of failure to work through this step.
8. *Generalize to other situations.* This may or may not be a necessary step, but it involves exploring the effect that the decision and its implementation may have on situations other than the immediate one.

9. *Evaluate the implementation.* This is a crucial step for determining whether the implementation plan and the decision choice were satisfactory. Too often we call a choice poor when in fact it was the implementation that was lacking.

Helpers can facilitate the preceding process by clarifying, providing information, and suggesting alternatives in the brainstorming step. In instances such as vocational and educational planning, information from tests may be used in the decision-making process. The gathering and synthesizing of pertinent information provide a valuable tool in cognitive decision making.

In addition to test interpretation and dissemination of appropriate information, helpers can use value clarification exercises, observation, and didactic teaching to aid helpees in learning to understand and to apply data obtained from tests, written and verbal information, and observation. These data in turn can aid helpees in clarifying and explaining their values, attitudes, and beliefs as well as their assets and liabilities. It is the helpee who makes the final decision; the helper provides invaluable assistance.

The following excerpt gives an example of decision-making techniques used in cognitive strategy.

Client: I'm having a hard time figuring out how to deal with the tardiness in my department.

Helper: It's frustrating, isn't it? Sounds like you're being held accountable for it.

Client: Oh, yes. As department head, I get landed on when the boss calls up in the morning and the girls aren't there to answer the phones and give him the information he wants.

Helper: So it becomes your problem. What kinds of things have you thought about doing?

Client: Oh, I don't know . . . docking people's pay, sending notes up to Personnel to be put in their files, making people make up lost time, rescheduling.

Helper: Sounds like you have thought of some options. Let's jot them down. Can you think of any others?

Client: I don't know. I suppose I could just ignore it and see what happens.

Helper: Have you thought about holding a department meeting and discussing it there? Maybe you could get some more ideas from your staff.

Client: Hmm . . . I don't know. We never seem to get anywhere when we discuss these kinds of issues.

Helper: Hmm. Maybe if you told them that your problem is that you're being held responsible for their tardiness . . . . Anything else we can put down as possible options?

Client: Wait a minute. I think before I really can consider different possibilities, I ought to see what my own staff has to say. Sort of put the shoe where it fits.

In this particular excerpt the decision-making process is not completed; rather, the helpee decides to get some more information before making a final decision. This is indeed appropriate, as sometimes decisions are made too quickly, before all the necessary data are collected.

When to Use Cognitive Strategies

Cognitive decision-making strategies are effective for educational and vocational planning and for problem solving and decision making in just about any life situation. Some categories of decisions require cognitive information and others need information about attitudes and beliefs.

The following exercises will help you to recognize how you make everyday decisions in your life and how other people can help.

Exercise 6.11   The purpose of this exercise is to put you in touch with your decision-making processes. Think back over the past 48 hours and write down every decision you made, such as what time you got up yesterday morning, what you had for breakfast, what you wore, and when you brushed your teeth. See how many decisions you made in that time span, and then go over your list and classify your decisions according to the following code: A = major decisions; B = commonplace, but not everyday, decisions; C = routine, everyday, taken-for-granted decisions. What does your list look like? Compare it with others'.

Exercise 6.12   In small groups, take a few minutes to jot down all the variables that influenced your decision to be where you are right now: in class, on a job, or wherever. See if you can go back to the very beginning of your decision-making process and identify a problem clarification point. Discuss your findings with your group and note similarities and differences.

Exercise 6.13   Divide into groups of six. Pick something that you all agree is a problem in your setting (for example, too much noise in the cafeteria or too few parking places). Try to decide how to solve the problem by going through the first seven steps of the decision-making process. Be sure you discuss each of the steps together.

☐ Cognitive-Behavioral Strategies

Cognitive-behavioral strategies are approaches that deal with both the thinking and the behaving processes. They are based on the premise that faulty thinking must be changed before effective behavior change can occur. The theoretical bases come from Ellis's rational-emotive therapy, Glasser's reality therapy, and Beck's cognitive-behavioral therapy, as well as from behavioral theory. Rationality and responsibility are key concepts in these approaches.

Techniques

Cognitive-behavioral techniques are largely verbal and require homework outside of the helping relationship to facilitate the transformation of new thinking into action or behavior.

The rational-emotive therapy model has contributed an effective strategy called cognitive restructuring, which means replacing faulty thinking with new, rational thinking. This strategy includes the didactic techniques of teaching, persuading, and confronting, and assigning homework. The purpose of cognitive restructuring is to aid helpees to control their emotions by teaching them more rational, less self-defeating ideas and convincing them of the illogic of the following irrational ideas identified by Albert Ellis (1962).

1. It is a dire necessity for me to be loved or approved of by everyone for everything I do.
2. Certain acts are wrong and evil, and people who perform these acts should be severely punished.
3. It is terrible, horrible, and catastrophic when things are not the way I would like them to be.
4. Much human unhappiness is externally caused and is forced on one by outside people and events.
5. If something is or may be dangerous or fearsome, I should be terribly concerned about it.
6. It is easier to avoid than to face life difficulties and self-responsibilities.
7. I need something other or stronger or greater than myself on which to rely.
8. I should be thoroughly competent, adequate, intelligent, and successful in all possible respects.
9. Because something once strongly affected my life, it should indefinitely affect it.
10. What other people do is vitally important to my existence, and I should make great efforts to change them in the direction I would like them to be changed.
11. Human happiness can be achieved by inertia and inaction.
12. I have virtually no control over my emotions, and I can't help feeling certain things.

The helper who uses cognitive-behavioral techniques continually unmasks helpees' faulty thinking by bringing it to their attention, showing them how irrational thoughts are the basis of their problems, demonstrating the A-B-C-D-E links (explained in the following paragraph), and teaching helpees how to rethink and reverbalize the preceding and similar sentences in a more logical, self-helping way. Thus helpers directly contradict and deny the faulty statements that helpees repeat to themselves, and they demand that helpees become involved in some kind of activity (homework) that will act as a "counterpropagandist" force against the faulty belief system.

In the A-B-C-D-E system, A stands for activating event, B stands for belief system, C stands for consequences, D stands for disputing irrational ideas, and E stands for new emotional consequence or effect. *The following excerpt demonstrates this system.*

Client: I'm really upset. I just did a job for a person and it didn't come out right. I can't stand it when things don't go right, and I just don't understand how it happened. It's really terrible!

Helper: Wow, you seem to believe that everything should always go right and that if it doesn't, you're no good. You're really doing a job on yourself.

Client: Well, I don't understand when things don't turn out right and I've done the right thing. I'm really upset.

Helper: I know you're upset. But you're upset because of what you're thinking about a job not coming out right, because you believe that everything should go right and if it doesn't, you're no good. Let me draw you something. See this A? That's the activating event, the job that didn't come out right. This B is your belief that everything should go right and that if it doesn't, you're no good. The C is the result of B, your feeling upset. Come on now, what kinds of things do you think you're telling yourself in B that are resulting in your feeling upset?

Client: I'm not sure. I guess I'm telling myself that I should always do well and that it's terrible when I'm not perfect.

Helper: That's really crazy. How about telling yourself that you did the best you could, that it's O.K. when everything does not always work out, and that you don't have to be perfect?

The homework in a case like this may consist of the helpee practicing new sentences, such as those suggested by the helper in the preceding example, every time he or she begins to feel upset when things don't go well. Helpees report back on their homework assignments. In conformance with Ellis's approach, the following kinds of rational ideas are taught:

1. It is not a dire necessity for one to receive love or approval from all significant others. One can concentrate on loving rather than on being loved.
2. It would be better not to determine self-worth according to external ideals of competence, adequacy, and achievement, but to focus on self-respect and winning approval for performance.
3. Wrongdoers ought not to be blamed or punished, but should be considered merely stupid or ignorant or emotionally disturbed.

4. One's unhappiness is caused or sustained by the view one takes of things rather than by the things themselves.
5. If something is dangerous, one should face it and try to make it non-dangerous, not make a catastrophe out of it.
6. The only way to solve difficult problems is to face them squarely.
7. It is usually far better to stand on one's own feet and gain faith in oneself and one's ability to meet difficult circumstances of living than to depend on someone else.
8. One should accept oneself as imperfect with general human limitations and specific fallibilities.
9. One should learn from one's past experiences but not be overly attached to or prejudiced by them.
10. Other people's deficiencies and weaknesses are largely their problems, and putting pressure on them to change is unlikely to help them do so.
11. People are usually happiest when they are actively and vitally absorbed in fulfilling pursuits outside themselves.
12. One has enormous control over one's emotions if one chooses to work at learning new, rational kinds of thoughts.

Needless to say, pointing out faulty thinking once is not likely to result in permanent behavior change. Rather, the helper must keep pounding away at the faulty belief system, time and time again. Helpers must insist on the completion of homework assignments that will demonstrate some behavior change.

The following is an excerpt from a rational-emotive therapy training session:
Mr. Whittier is 66 years old, retired, and widowed. He is currently living alone in his apartment of 40 years. He has three grown sons who live in other towns. He's been referred to a counselor because of continued moping and self-pity.

Mr. Whittier: I'm all alone now. The boys, they've gone off, and they don't really bother with me now. I suppose I should move to a smaller apartment, but all the memories are here. Where have all the years gone to?

Helper: Mr. Whittier, you seem to think that your boys should be more attentive to you.

Mr. Whittier: Yes . . . why not? What are children for? We all had some good years together.

Helper: It certainly would be nice if your boys did pay more attention to you, but the fact that they don't doesn't mean you have to stop living, you know.

Mr. Whittier: What do you mean?

Helper: You keep acting as if you can't live without them. You go around here telling everyone how terrible it is that your sons don't write more, don't visit and call more. But you are healthy and you are living, and things don't always have to be the way you want them to be.

Mr. Whittier: What kind of talk is this? Things *don't* always have to be the way I want them to be . . . um . . . let me think.

Helper: Isn't that what you keep on telling yourself?

Mr. Whittier: *(long pause)* Maybe, maybe, young lady, you have something.

Helper: Let me help you think it out. There are some new sentences you can learn to say over and over to yourself every time you feel yourself getting upset with your sons. If, instead of telling yourself over and over again how terrible it is that your boys don't call, write, and visit more, you could learn to say instead, it's too bad those boys don't call, visit, write more—we could have good times together—but I'm making friends of my own here at the center, and I'm going to manage and live anyway. . . .

In this case the helper had gotten to know Mr. Whittier over several years at the community center, so she felt comfortable confronting him with his irrational thinking.

Reality therapy uses a different technique in the cognitive-behavioral domain. Involvement between helper and helpee is crucial to reality-therapy techniques, which involve eight steps:

1. Get involved; be personal and communicate "I care about you" by words and actions.
2. Stay in the here and now; avoid references to the past and avoid dwelling on feelings. What the person does with them is more important than the feelings themselves.
3. Evaluate behavior; ask the helpee to evaluate his or her own behavior and ask, "Is what I did appropriate?" "Is it helping me . . . others?" If the client cannot evaluate his or her behavior to your satisfaction, it is necessary to return to step 1. The helpee must decide whether or not to change his or her behavior.
4. Plan to change behavior; ask "What do you think would be a better way to do things?" Help the client formulate a plan. Let the client choose; the helper offers suggestions but does not provide the plan. The plan should be minimal, specific ("How and when will you do this?"), positive rather than negative or punitive, and have a high probability of success.
5. Contract to seal the plan. If necessary, write out a contract and have the helpee sign it. Follow through with a check on how it is going and support success. This contract is between the helper and the helpee.
6. Accept no excuses for failure to fulfill the plan. If the contract is not fulfilled, ask "When will you do this?" not "Why didn't you do this?" If unsuccessful, follow through with the natural consequences of not having followed the plan, and then go back and make a new plan.

7. The helpee should know and be involved with making the rules. Use natural consequences (results when the rules are broken) rather than punishment.
8. Never give up.

A helper using reality-therapy techniques will become very involved with the helpee, who can then begin to evaluate his or her own behavior and see what is unrealistic. Helpers confront clients with reality and ask them again and again to decide whether or not they wish to take the responsible path. Helpers then ask clients to make specific plans and to take responsibility for implementing them. Helpers can reject unrealistic behavior but still accept helpees and maintain respect for them. Helpers teach helpees better ways to fulfill their needs without hurting themselves and others. Helpees assume responsibility for their behavior, work in the present, learn to assess the morality of their behavior, and learn more effective ways of behaving.

An example of this approach follows.

Client: I don't want to finish school. I hate the teachers, I can't learn anything from them, and it's no use going back there.

Helper: What's going on in school? What are you doing?

Client: I'm always getting sent to the office. My English teacher is a real nag. She picks on me for everything, and I know she has it in for me.

Helper: What are *you* doing?

Client: Nothing much. I just sit there . . . sometimes I mess around a little.

Helper: What's "messing around"?

Client: Oh, you know, talking to other guys, fooling around . . . .

Helper: Do you think this is appropriate behavior?

Client: Aw, I don't know. Most of the guys do it.

Helper: Let me ask you something. If you don't go back to school, what will you do?

Client: I'll get a job. I want to go up north and work at a ski resort. I can be an instructor.

Helper: You think you can get that kind of job without a high school diploma?

Client: I think so . . . I don't know. Why would a ski instructor need a high school diploma?

Helper: Well, it might help. It will certainly give you more choices for the rest of your life.

Client: I'm sick of school. I really don't want to go back.

Helper: It's up to you to decide what you want to do. I think you should think about what you're doing and whether or not it is going to get you what you want in this world.

This excerpt shows the implementation of steps 2 and 3 (step 1 had been started in previous sessions). As previously mentioned, reality therapy uses a teaching strategy that deals directly with choices of the helpee. The basic philosophy is that the helpee can decide whether or not to be troubled.

Beck's form of cognitive-behavioral therapy uses a wide range of core strategies incorporating cognitive and behavioral techniques. Many of these resemble Ellis's cognitive restructuring. Some of Beck's strategies include cognitive rehearsal to identify roadblocks in thoughts, associating feelings with behaviors by imagining situations in every detail during the session, many types of reality testing such as finding alternative responses to negative thoughts, task assignment, and actively testing out negative thoughts and assumptions.

Helping the client to become aware of and distanced from faulty thinking may prevent similar future errors. Beck (1976) outlines seven steps of a reality-testing technique that illustrates his application of strategy.

1. Identify thoughts and statements made by client that are negative or associated with bad feelings.
2. Ask client how much he or she believes that the statement or thought is true or how likely it is that the negative event will occur.
3. Check feelings associated with the statement—for example, "When you say that to yourself, what does it make you feel?"
4. Leaving the validity of the statement as an open question, gently probe the evidence: past outcomes of similar situations, alternative outcomes and their frequency, times when the same situation has had better or worse consequences than presently imagined, and so on.
5. Rate possibility of catastrophes in the future—for example, "How probable is it that you will never find another friend like him? One in ten? One in one hundred?"
6. Continually challenge thoughts with reality.
7. Check how much client believes original statement is true after going through these steps.

Note that cognitive-behavioral techniques include evaluation and judgment on the part of the helper, who labels the helpee's thinking and behavior as rational or irrational, responsible or irresponsible. Helpers do not arbitrarily impose their value systems on helpees; rather, they examine and evaluate the helpee's values. In other words, helpers challenge helpees but do not punish them or reject them for not having the "right" values or beliefs. This approach does differ, however, from the phenomenological strategies, which are both nonjudgmental and nonevaluative.

When to Use Cognitive-Behavioral Strategies

These approaches have been used with a wide variety of the population, in schools, hospitals, industry, and correctional institutions. The rational-emotive therapy approach might not be effective with helpees who are not intelligent enough to follow a rational analysis or for those who are so caught up in emotion that they cannot attend to this logical procedure. The reality-therapy approach involves much straightforward common sense and has also been used with a broad spectrum of the population. Beck's cognitive-behavioral therapy has been particularly effective with depressed clients and is now applied to a wide array of disorders. This therapy requires verbal ability and the motivation for change on the part of the client, as do other cognitive approaches.

The following exercises will provide you with opportunities to question your own belief system and to design a reality-therapy contract.

Exercise 6.14   In triads, share the last strong negative feeling each of you has experienced. Describe the circumstances, and help each other to analyze irrational thinking via the A-B-C-D-E system. Which of the 12 irrational ideas does the faulty belief system come from? What kinds of new sentences can you teach yourself to refute the old sentences you've been telling yourself? Prepare a homework assignment for yourself and report back to your triad in one week.

Exercise 6.15   In groups of six, reverse roles such as masculine/feminine, supervisor/supervisee, and black/white, depending on the makeup of your group. Role play a scene and then discuss the assumptions and conceptions you've become aware of while role playing an unfamiliar role and observing others playing a role familiar to you. What irrational beliefs have you uncovered? What do they mean to you?

Exercise 6.16   In small groups, construct a reality-therapy contract for at least one member of the group in accordance with the eight steps previously discussed. Stick to specific behaviors and identify the logical consequences of meeting and not meeting the contract. What did you find functional or dysfunctional for you in this process? How well were you able to identify responsible and/or irresponsible behavior?

Exercise 6.17   Imagine being in a situation that would probably upset you. Perhaps you can think of something that has actually occurred in the past week or is likely to occur in the next few days. Imagine that you are experiencing this right now, with its whole range of accompanying negative thoughts and feelings. Now tell a partner what the situation is and share those negative thoughts and feelings. Together, attempt to brainstorm resolutions for the situation. During or after generating some options, develop each possible course of action in great detail to discover possible roadblocks.

☐ Behavioral Strategies

Behavioral strategies are based on learning theory and focus on specific, observable behaviors as opposed to feelings and thoughts. The objectives of these strategies are to alter inappropriate behaviors and to teach appropriate ones. The assumption is that change in behavior results in changes in feelings and thoughts and that helpers can evaluate their effectiveness only by observing concrete, specific behavior changes.

Techniques

The many behavioral techniques require some skills of the helper. Some of those skills are the following:

1. Understanding of the concepts and principles of reinforcement, punishment, extinction, discrimination, shaping, successive approximations, and schedules of reinforcement
2. Ability to identify specific target behaviors that the helpee wishes to change
3. Ability to identify and assess the conditions preceding the helpee's target behavior
4. Ability to collect baseline data on the frequency and severity of target behavior
5. Ability to identify and assess those conditions that both result from the target behavior and maintain (reinforce) them
6. Ability to determine reinforcements that are meaningful for the helpee
7. Ability to determine feasible and meaningful schedules of reinforcement
8. Sufficient knowledge of the theoretical framework, design, and application of different behavioral strategies
9. Ability to evaluate outcomes of behavioral strategies

It is beyond the scope of this book to teach you these skills, but you will find readings at the end of this chapter that cover this material. Behavioral strategies are being taught to teachers and parents and are increasingly being used in businesses, schools, and health organizations. Many paraprofessionals and beginning professionals are assisting in the implementation of these strategies. Some important behavioral techniques that we will discuss are modeling, contracting, **assertiveness training**, and systematic desensitization.

Modeling is based on the principle that people learn to behave in new ways by imitating the behavior, values, attitudes, and beliefs of significant others. Modeling may be accomplished through role playing, the use of media, and individual and group counseling relationships. Remember that the helper is a model, a very powerful model, in a helping relationship.

An example of role-play modeling follows.

Client: I always have a hard time at those dorm parties. I'm never able to go up and start a conversation with someone I don't know.

Helper: Um. Let's role play. I'll be a stranger at a party and you be standing by the wall with a beer.

Client: O.K. *(rearranges self)* Ummmmm . . . hot, isn't it?

Helper: Yeh.

Client: Lots of people here. . . . Oh darn! See, I can't do it; it just goes nowhere.

Helper: O.K. Now let's reverse roles and see what happens. Hi, there. Some crowd here . . . it's difficult getting around. You struck me as someone interesting to talk to.

Client: I did? Oh, well, thanks. That's nice.

Helper: I was wondering what you're thinking about all this *(gestures around the room)*.

Client: I see what you did. You asked broader questions and I asked dumb ones that didn't need answering.

Helper: That's one thing. What were you feeling when we did this?

In modeling, it's important for the helper to be aware of his or her modeling influence on the helpee and of positive and negative models in other aspects of the helpee's life. Sometimes helpers arrange for people to work or study together or in small groups, having definite models in mind. By the same token, helpers sometimes break up existing groups or cliques because of the effect of negative models.

Contracting is based on theories of reinforcement, which state that reinforced (rewarded) behavior tends to be repeated. A behavioral contract is a specific agreement between helper and helpee that breaks down the target behavior into its smallest parts and provides for systematic reinforcements of the performance of the behavior.

Contracts may be informal (for example, "If you do X, I will do Y") or formal (written statements of specific behavior to be performed, the specific reward to be granted, and the specific responsibilities and conditions for implementing and monitoring the contract).

Contracts should still follow the basic rules proposed years ago by Homme (1970) in that they should (1) use reward liberally and immediately following performance, (2) be clearly understood by all parties, and (3) be expressed in positive terms, meaning that they should state what one is to do, not what one is not to do.

Here is an example of a formal contract drawn up between a residence advisor, a college student, and the student's girlfriend in a dormitory. The problem was that the student was in danger of failing a chemistry course because he was falling behind with homework assignments and getting low test scores. Tom, Mary Beth, and Len agreed to participate on these terms.

If Tom reads one chapter of his chemistry text and completes the problems at the end of the chapter with 75 percent accuracy by 9:30 each evening, he will meet

Mary Beth in the lounge for coffee between 10 and 11 P.M. Len will be available to correct the problems at 9:30 P.M. If Tom does not meet these terms, he will stay in his room the rest of the evening and not meet with Mary Beth. This contract will be reviewed after two weeks.

As in reality-therapy contracts, there is no punishment or rejection for an individual not meeting the terms of the contract; however, reinforcement is given only after performance of the target behavior, and it is important that this reinforcement not be available outside the terms of the contract. Contracts must be positive and within the realm of possibility. The target behavior is broken down into its smallest component parts, and an appropriate amount of reinforcement is administered for each small component behavior performed. For example, Tom may be far enough behind in his chemistry to warrant reading two chapters per night, but the contract begins with one chapter because Len and Tom know that Tom can definitely meet those terms. When the contract is reviewed, the terms may change.

The use of contracts in a helping relationship has the advantage of specifying in positive terms exactly what is expected from the helpees and what they will receive for fulfilling those expectations. Contracts enable some movement and growth in problem areas, resulting in higher self-esteem and thus permitting attention to be turned to other areas of concern. Contracts are ethical as long as they do not specify performance of immoral behavior and as long as all involved parties agree to the terms.

Assertiveness training is used in the cognitive as well as the behavioral domain. It can involve changing helpees' belief systems by teaching them that they should stand up for their own rights as long as they do not harm someone else or impinge on someone else's rights in the process. This kind of training reduces helpees' anxiety by teaching them to say what they want to say. Methods of assertiveness training can involve role playing **successive approximations** of the desired responses, a little bit at a time, leading to a totally assertive response, modeling from media, and verbal instructions and illustrations.

Here is an example of assertiveness training for a 25-year-old divorcée.

Client: I've been in a terrible state all day. Since I left Russ, my parents have really been on my back. My dad called last night and announced they're coming up this weekend, and now I have to change all my plans. I've been unable to concentrate on anything all day.

Helper: Sounds to me like you're really uptight about this. You were unable to tell your parents that this weekend was inconvenient for you.

Client: Yes. I never can tell them what I feel. They get so upset and carry on so. But I hate myself for getting into this state.

Helper: I wonder what would happen if you called your dad tonight and told him you'd like to see him, but this weekend is inconvenient for you and you'll let him know when he can come up.

Client: I wish I could do that. I'd love it. I don't think I can.
Helper: Let's rehearse and see what happens.

They then role played the same scene over and over, with the helper playing both roles at different times for modeling purposes. After the fifth rehearsal, the client reported that she felt less anxious. The next day she reported that she had phoned her father and felt relieved that she had been able to take this step. Her assertive responses did not contain a "You're no good, you have no right to interfere in my life" kind of message. She sent an "I" message, communicating her interest in seeing her father, but letting him know, clearly and firmly, that she had her own life to lead and would let him know when it was all right for him to come up. As she continues to practice this kind of behavior, she will come to change her belief system and really believe that she has the right to her own life.

Systematic desensitization involves breaking down anxiety-response behaviors and exposing the helpee to the imagery of those behaviors while in a state of deep physical relaxation. The theory is that anxiety responses have been **conditioned** (learned) and can be counterconditioned (unlearned). One way to countercondition anxiety responses is to pair them with an incompatible state—in this case, a physiological state of relaxation, which inhibits anxiety. The anxiety stimulus eventually loses its potency, and the helpee no longer needs to expend energy on the anxiety response.

It is unlikely that many human services workers will actually apply complete systematic desensitization; but the first stage, that of inducing deep muscle relaxation, is often helpful in and of itself and is easy to learn to apply. It usually takes two to three sessions to effectively teach relaxation. Between sessions, helpees are asked to practice relaxation skills at least once a day. Helpees are taught to contract (tense) and then relax specific muscle groups for several minutes at a time until they learn to monitor their own relaxation.

Before beginning this training, the helper explains the process to the helpee. The helper explains that it is impossible to be both tense and relaxed at the same time and that, once helpees learn relaxation, they can use this training whenever they are tense. This is skill learning and requires continual practice in order to be effective.

At the beginning of the training session, the helpee is seated in a comfortable chair that supports all body muscles. Eyes are closed (the helpee is asked to remove eyeglasses or contact lenses), the head is supported by the chair or wall, arms are on the armrests of the chair, and legs are uncrossed and firmly on the floor. Relaxation training commences after the helper demonstrates to the helpee the tensing and relaxing of each muscle group. It is a good idea to darken the room and reduce interfering noises.

The following is an example of relaxation instructions. (I am indebted to Flora Hummel, R.N., for introducing me to this technique.)

Close your eyes now, lean back, and just get comfortable. Think about your body and what you're feeling now . . . now I want you to raise your hands and clench your fists as tightly as you can . . . feel those muscles pulling . . . tighter now . . . O.K. 1 . . . 2 . . . 3 . . . 4 . . . now relax and let your hands fall into your lap. Now let's do that again . . . tighten those hands . . . 1 . . . 2 . . . 3 . . . 4 . . . now let them relax again . . . now raise your forearms and pull those muscles as tightly as you can . . . 1 . . . 2 . . . 3 . . . 4 . . . relax and let them fall into your lap . . . attend to the different feelings you get from relaxed and tense muscles . . . now let's do those muscles again . . . 1 . . . 2 . . . 3 . . . 4 . . . relax . . . feel your arms getting heavier as they become more relaxed . . . now pull in on your upper arm muscles by raising your arms and flexing those muscles . . . harder, now . . . that's right . . . 1 . . . 2 . . . 3 . . . 4 . . . relax . . . your arms feel heavier and warm feelings are spreading down through your fingertips . . . now let's do that again . . . 1 . . . 2 . . . 3 . . . 4 . . . relax . . . we'll concentrate on your head muscles now . . . let your arms continue to grow heavier and relax more while you think about your head muscles . . . now raise your eyebrows and wrinkle up your forehead as much as you can . . . tighter . . . hold it . . . 1 . . . 2 . . . 3 . . . 4 . . . now relax, and feel that tension slipping out over your head, right over the top . . . let's do that again . . . hold it . . . 1 . . . 2 . . . 3 . . . 4 . . . relax . . . now scrunch up your eyes and feel them get as tight as possible . . . 1 . . . 2 . . . 3 . . . 4 . . . now do that again . . . 1 . . . 2 . . . 3 . . . 4 . . . relax and feel your eyelids getting heavier as everything gets darker . . . now scrunch up your nose as tight as you can . . . hold it . . . 1 . . . 2 . . . 3 . . . 4 . . . now relax . . . do it again . . . 1 . . . 2 . . . 3 . . . 4 . . . that's right, now relax and feel your breathing get clearer and easier . . . now stretch your mouth in an ear-to-ear grin and pull on your lip, jaw, and cheek muscles . . . come on, you can pull tighter than that . . . hold it . . . 1 . . . 2 . . . 3 . . . 4 . . . now relax . . . let your jaw hang loosely and your head lean right into the back of the chair . . . do that again . . . 1 . . . 2 . . . 3 . . . 4 . . . that's right, now relax . . . now think back over your arm muscles and let your head rest heavier and heavier . . . you're getting more and more relaxed and all you're thinking about is your muscles getting more and more relaxed . . . now pull in your neck and throat muscles . . . feel that tension . . . hold it . . . 1 . . . 2 . . . 3 . . . 4 . . . now pull in your neck and throat muscles . . . feel that tension . . . . hold it . . . 1 . . . 2 . . . 3 . . . 4 . . . relax . . . let your neck and head slump into the chair . . . how about repeating that . . . 1 . . . 2 . . . 3 . . . 4 . . . relax and feel your neck get looser and looser . . . now raise your shoulders and pull those muscles tightly . . . hold it . . . 1 . . . 2 . . . 3 . . . 4 . . . that's right, now relax and let those shoulders slump . . . let those warm, tingly, relaxed feelings connect up from your shoulders down your arms . . . that's right . . . repeat that . . . 1 . . . 2 . . . 3 . . . 4 . . . relax . . . feel those relaxed feelings getting deeper and deeper . . . now tighten up your upper back muscles by arching it as much as you can . . . 1 . . . 2 . . . 3 . . . 4 . . . relax . . . slump down into your chair . . . do it again and tighten your chest muscles at the same time . . . 1 . . . 2 . . . 3 . . . 4 . . . relax . . . take a couple of minutes to feel more relaxed and go over your arm muscles and your head muscles . . . if any group of muscles starts to tighten up, pull them in, contract them and then relax them, like a rubber band . . . now pull your tummy muscles in as tightly as you can and feel that tension . . . hold them . . . 1 . . . 2 . . . 3 . . . 4 . . . relax . . . feel your stomach get looser . . . repeat that . . . 1 . . . 2 . . . 3 . . . 4 . . . relax . . . breathe deeply and slowly and with each breath in, feel more relaxed and with each breath out, feel that relaxation spreading throughout your body ... now tighten your buttocks and hold them tight ... 1 ... 2 ... 3 ... 4 ... relax ... now do that again ... 1 ... 2 ... 3 ... 4 ... relax and sink deeper and deeper into the chair ... now we'll work on your leg muscles ... tighten your thigh muscles ... 1 ... 2 ... 3 ... 4 ... relax ... do that again ... 1 ... 2 ... 3 ... 4 ... feel your legs getting heavier and the warm, relaxed feelings spreading down ... now tighten up your calves by pointing your toes away from your head ... 1 ... 2 ... 3 ... 4 ... relax ... repeat ... 1 ... 2 ... 3 ... 4 ... that's right ... the warm tingly feelings are spreading down your legs ... now crunch up your toes and tighten those feet muscles ... 1 ... 2 ... 3 ... 4 ... relax ... now do that again ... 1 ... 2 ... 3 ... 4 ... fine ... now think back over all the muscle groups we've relaxed and see if they can become more and more relaxed ... that's right ... the arms ... your head ... your shoulders ... back ... stomach ... buttocks ... legs ... take a few minutes to really get in touch with these warm, tingly feelings ... now I'm going to count to 20 very slowly, and when I use an odd number, breathe in, and when I use an even one, breathe out (*counts to 20 very slowly*).

At this point either desensitization begins or, if relaxation is being used alone, after several moments, the helper counts slowly to ten and asks the helpee to open his or her eyes and slowly stretch.

If desensitization follows, it is in this relaxed state that the helper asks the helpee to call different scenes to mind: a neutral scene (like a blank screen), which does not arouse any feeling; a comfortable scene (like sitting in front of a fireplace or by the ocean), which arouses only comfortable feelings; or a scene that can arouse anxiety. Anxiety scenes are first introduced for 10 seconds and then followed by a neutral or pleasurable scene and more relaxation concentration; then they are introduced for 20 seconds, and again, after more pleasurable scenes, for 40 seconds. If anxiety occurs, the helpee signals the helper, and the length of the anxiety scenes is reduced. An anxiety item is considered successfully mastered when the subject can imagine it for three 40-second periods without experiencing any anxiety.

A desensitization hierarchy is constructed during the first two relaxation-training sessions. This hierarchy comprises a series of statements that describe anxiety-provoking stimuli, ranked from least to most anxiety-provoking.

A sample of a hierarchy used for test desensitization follows (stimulus number 1 causes the least anxiety; number 13, the most).

1. Walking to class.
2. Professor announces examination is two weeks away.
3. Copying classmate's notes for missed class.
4. Obtaining references to study.
5. Discussing class with friends.
6. Studying the week of the exam.
7. Studying the night before the exam.
8. Waking up the morning of the exam.
9. Walking to the classroom.
10. The exam is being handed out—you receive a copy.

11. While trying to think of the answer to an exam question you notice everyone around you writing quickly.
12. You come to a question you can't answer.
13. Professor announces that 40 minutes remain; you have one and a half hours' work to do.

When the cause of anxiety is something that can actually be tested in the helping relationship, you can validate the client's desensitization to it. For example, I once treated a client who had a driving phobia. After reducing her anxiety through systematic desensitization, we entered my car and found that she was able to drive for the first time in 18 years.

When to Use Behavioral Strategies

Behavioral strategies are effective with a wide population and have proven to be especially effective with people who have difficulty with verbal strategies. These action strategies are usually shorter-term than some other types of strategies.

In particular, modeling is effective for those who are unsure of themselves and need specific teaching examples. Contracting is helpful for retarded as well as "normal" populations and is particularly effective in families and organizations where reinforcements can be immediately provided and monitored. Assertiveness training is helpful for shy, inhibited people. This technique has also been used in women's consciousness-raising groups. Systematic desensitization is helpful for those with phobic reactions such as fear of flying and fear of water.

The following exercises will give you some experience with the previously described behavioral techniques.

Exercise 6.18   The purpose of this exercise is to check your ability to identify behavioral statements (statements that describe behavior that is observable and measurable). Which of the following statements are behavioral? (Answers are given at the end of the chapter.)

1. "That child is no good—he's stubborn and fresh."
2. "Johnny never sits down at his desk. He's always up walking around."
3. "I feel so guilty when I think of him all alone."
4. "Ms. Leonard came late to the board meeting."
5. "Ms. Leonard really isn't interested in this organization."
6. "Pam is so spoiled—she whines and complains a lot."
7. "My husband doesn't appreciate me."
8. "He's lazy, just like his father."
9. "He never seems to get anything done on time."
10. "I had a boring day today."
11. "I cleaned the house all day."
12. "She really is a good mother."

13. "My secretary is the fastest typist I've ever seen."
14. "He belongs to a gang."
15. "Gangs are roaming the streets at night."

Exercise 6.19   Individually, write down five behaviors that you think an effective helper ought to exercise. Then, in small groups, agree on one list. Then each group can share its list. Do the lists all contain behaviors or do some of the items represent attitudes and evaluations?

Exercise 6.20   *Modeling:* Who are the most influential role models in your life? What are their most important characteristics, in your opinion? Discuss these questions in small groups and see if you can draw up a list of effective model characteristics. Compare lists among the small groups.

Exercise 6.21   *Reinforcement:* In triads or small groups, discuss what is reinforcing to each of you in your life. See if you can identify the major social (from people, such as a smile, gesture, or visit) and concrete (things, such as money, gifts, or purchases) reinforcers that operate in your life at home, at work, and at leisure. See if they differ in the different settings. As you discuss each other's reinforcers, you will see that different people have different kinds of reinforcers, that what is reinforcing for one person may or may not be so for another.

Exercise 6.22   *Reinforcement:* Imagine yourself in a familiar setting, either at home, work, or school, and see if you can determine what kinds of reinforcements, both social and concrete, you give others and yourself. For example, "When I am at home and complete a housekeeping task I dislike, I usually reward myself by talking to a friend on the phone or having a cup of coffee with a neighbor." Share your responses in small groups.

Exercise 6.23   *Contracting:* In pairs, draw up contracts for each of you, listing your partner as monitor. This contract may be between you and your partner or between you and someone else. Select a target behavior (like losing so many pounds or getting to class on time), determine a time limit, and write out what you will do, when you will do it, and what you will receive for your performance. For example, "If I lose two pounds between today and next Thursday, I will buy myself a new shirt." It is essential that the reinforcement be something that is meaningful to the helpee, not to the helper.

Exercise 6.24   *Assertiveness training:* In each of the following situations, which response is the most assertive? (Answers are at the end of the chapter.)

1. Your mother telephones you long distance and wants to know why she hasn't received a letter from you all week.
   a. "It's hard for you to realize that I'm grown up now. I'll write when I can."
   b. "I'm sorry, Ma, I've just been too busy; I'll write tonight."

c. "Oh, Mom, please stop bugging me. I'm not a baby, you know."

2. A co-worker asks you to get him coffee. This is the umpteenth time, and you do not want to do it.
   a. "I guess so, since I'm going down there anyway."
   b. "Why can't you get your own coffee?"
   c. "I'd really feel better about our relationship if you would stop asking me to get your coffee."

3. Someone pushes in front of you at the box office line.
   a. "Who do you think you are?"
   b. "I'm ahead of you in this line. Please move."
   c. "Some people really have a nerve!"

4. Your boss asks you at the last minute to stay late to work on a report. You know it is not an emergency, and this is the third time this has happened this month.
   a. "O.K. You're the boss."
   b. "But we're having company tonight, and I promised Marge I'd pick up some ice."
   c. "I do have other plans tonight, and I'm afraid it's too late for me to change them."

5. Your friend wants to borrow your car. Last time this happened, you promised your husband you wouldn't do it again.
   a. "I wish I could, but Jim made me promise not to lend out the car."
   b. "I'd like to help out, but Jim and I have agreed that we can be more accountable about the car if we don't let others drive it."
   c. "Oh dear, I think I'll need it when you want it."

6. The sales clerk tells you that the store does not accept returns or exchanges. If the merchandise is defective, you'll have to deal directly with the manufacturer.
   a. "That's ridiculous. I'm going to call the Better Business Bureau."
   b. "I insist that you refund my money, and I'm not budging until you do."
   c. "I'd like to see your manager, please. I intend to straighten this out here and now."

7. Your doctor tells you that there is nothing wrong with you, but he would like you to have some tests made anyway.
   a. "Before I have any tests, I'd like to know the reasons, costs, and just what's involved."
   b. "But why do I need these tests if there's nothing wrong with me?"
   c. "Is this another example of useless tests I keep reading about?"

8. Your friend pleads with you to go out to dinner, and you want to stay home alone and relax after a hard week.
   a. "I have a headache, so I want to go to bed early."
   b. "I'd really like to be alone tonight and relax. Some other time."
   c. "Well, why don't you come over here, and we'll fix something to eat?"

9. A neighbor calls you to complain about your son being a bully. This has happened before with this particular neighbor, and you have checked it out and determined that it is your neighbor's problem, not yours.
   a. "I'm sorry. Thanks for letting me know about this. I'll talk to him."
   b. "You really ought to find out about your own kids before you complain about mine."
   c. "I'd appreciate it if you'd check out both sides of the situation, as I will."
10. You've been sitting in the restaurant for almost an hour and you still have not been served, even though you told the waiter you had a curtain time to make.
   a. "Waiter! What's taking so long? I told you I have to be at the theater in half an hour."
   b. "Waiter! Curtain time is in 30 minutes. I expect to have finished eating by then."
   c. "Waiter! Will you please hurry? The service around here is terrible."

Exercise 6.25   *Assertiveness training*: Arrange a line of seven or eight people, such as one sees at a grocery checkout counter, at the box office of a movie, or waiting to get on a bus. Those in your group who are not in line should try, one at a time, to break in and push ahead. See how each of you reacts and then discuss those reactions in your group. The purpose of these last two exercises is to help you become aware of your own assertive behaviors or lack of them. In smaller groups, identify situations that require assertive (not aggressive) behaviors, and then role play those situations in as many ways as you can.

Exercise 6.26   *Relaxation*: Using the relaxation-training example in this chapter, see if you can help one or more persons in your group to relax. Each time a set of ellipses appears in the example, allow several seconds to elapse. Keep your voice soft and soothing, and be careful not to rush. Experience the role of trainee as well as trainer, and share your feelings and reactions.

☐ Strategies That Cut across Domains

Cutting across the affective, cognitive, and behavioral domains are transactional analysis (TA), multimodal therapy, systems strategies, and other eclectic approaches. These draw on many of the theoretical bases of the other approaches.

Eclectic strategies can be used when the helpee's problem does not fall clearly into one domain, when it overlaps into two or three domains. For example, one may use a behavioral strategy for modifying a particular behavior and, at the same time, attempt some cognitive restructuring (to change the client's thinking about the target behavior) and/or elicit the client's feelings by using affective strategies.

Eclectic therapists believe the more domains addressed, the stronger the likelihood of change in the helpee.

The multimodal BASIC ID model discussed in Chapter 5 provides a specific, functional basis for the selection of an eclectic use of techniques. For example, clients who are restricted in awareness and expression of their feelings may benefit from Gestalt as well as client-centered methods. Clients who are well aware of their thoughts and feelings might benefit by behavioral change techniques. The BASIC ID paradigm allows helpers to prioritize helpee needs and goals. Ivey's DCT model uses strategies drawn from psychodynamic, client-centered, behavioral, cognitive, and systems theories. Microskills such as responsive listening and questioning are the helper's core tools. These strategies are particularly useful for life-span development issues across cultures.

TA Techniques

TA techniques include analysis of games and scripts, identification of ego states, and analysis of communication patterns and transactions. Gestalt experiments and games are used to supplement those techniques. TA is a contractual and reinforcement strategy. The vocabulary and concepts of TA must be mastered before the techniques can be used, so helpers first explain them to the helpee and usually assign some reading material.

The following is an excerpt from a TA counseling session.

Client: I'm unhappy in my marriage. My husband is always picking on me and nagging me. I feel so low.

Helper: Stay with that low feeling. Your voice has dropped, too. You sound like a scared little girl.

Client: I guess I am scared. It just seems as if all we do is fight these days.

Helper: Your husband is relating to you out of his critical parent state, and you're responding by feeling low and scared in your child ego state. What does that mean to you? Sounds like you're getting something out of it.

Client: I guess I at least get his attention that way. It's just that I never seem to do anything right for him. The meat is too well done, the laundry isn't picked up on time, the kids aren't well behaved.

Helper: This scared feeling is familiar to you.

Client: Yes, all my life I've been scared.

Helper: I have a hunch that when you were a kid you were told you could never do anything right.

Client: My mother was always criticizing me. How did you know that?

Helper: So you learned how to be scared and play "stupid." You and your husband have a real game going there. It looks something like this. *(Client is referred to a drawing similar to the one in Figure 6.1.)*

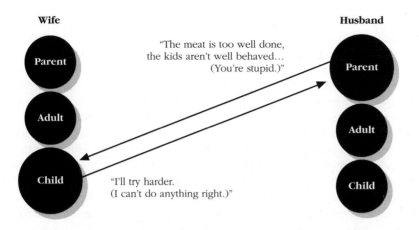

Figure 6.1 Diagram of a transactional analysis game

The preceding excerpt shows the identification of ego states and the "racket" of being scared, and is just getting into identification of the game. The husband and wife are involved in a parent-child game in which the husband occupies an "I'm O.K., you're not O.K." position and the wife occupies an "I'm not O.K., you're O.K." position. The helpee is just beginning to explore the early childhood decision that resulted in the script to do everything wrong and feel "not O.K." about it.

Helpers use all three of their ego states to help clients activate their adult ego state as the executive of the personality. Helpers use their nurturing parent ego state to provide protection, caring, and permission; their adult ego state to provide skill and expertise in helping; and their child ego state to provide fun and intuition.

A major technique of TA is called *reparenting*, which occurs when the old, critical parent ego state is replaced with a new, nurturing ego state that allows for rescripting (changing those early childhood decisions). One way helpers can identify ego states is to have the helpee, by means of a Gestalt dialogue, role play the different ego states interacting with one another.

When to Use TA Strategies

TA is particularly effective for persons with interpersonal difficulties and for persons who seem to function inadequately (perhaps because of ego state **exclusion** or **contamination**). TA materials that can be used with young children are included in the reference list at the end of this chapter. Because the helpee must be able to understand the vocabulary and concepts of TA, it would probably not be effective with retarded populations.

The following exercises illustrate some of the TA principles.

Exercise 6.27   In small groups, role play dialogues and see if group members can identify which ego state is operating in each participant. Note tone of voice, choice of words, body posture, facial expression, gestures, and so on. The nurturing parent ego state is loving, caring, supportive, and rescuing and consists of caresses, tender smiles, and loving words. The critical parent ego state includes values, prescriptions, injunctions, frowns, pointing finger, punishments, and words such as "do," "don't," "should," and "ought to." The adult ego state is like a computer; it provides information, asks "what," "when," and "where" questions, lacks feelings, and is entirely objective, with a level, clear tone of voice. The child ego state is spontaneous, intuitive, angry, defiant, and complacent. It giggles, cries, scowls, gestures obscenely, sulks, and uses words like "golly," "gee whiz," "I can't," and "oh, boy." Suggestions for scenes are (1) a dialogue between parents at the dinner table about their 17-year-old daughter's hours, (2) a dialogue between physician and patient, (3) a dialogue between welfare worker and client, and (4) a dialogue between boss and employee.

Exercise 6.28   In small groups, see if you can identify games that each of you has played. Games are "unstraight" ways that your "not-O.K." child uses to get stroked (reinforced) and avoid responsibilities. Common games are "Stupid," "If it weren't for you," "Kick me," and "Harried." What kinds of games do students play? Teachers? Parents? Helpers?

Exercise 6.29   Brainstorm, in small groups, ways that you can make your home, work setting, or the setting you are now in "more O.K." How can you give yourself and others more positive strokes? Each person in the group should share three things he or she likes about himself or herself, and then each person can take responsibility for giving at least one stroke (positive only!) to one person in the group, based on what was previously disclosed. The purpose of this exercise is to learn how to give and receive strokes, to take responsibility for what makes you feel O.K.

Exercise 6.30   The purpose of this exercise is to help you get in touch with your script. To do this, we help you recollect early decisions you made that influenced the development of your script. Answer the following questions and see what you can derive from those answers.

1. Who was your most significant parent? (Name the one who pops into your head first!)
2. What was the best thing that parent said to you as you were growing up?
3. What was the worst thing that parent said to you as you were growing up?
4. How did you feel about the best thing and the worst thing? What did you think about each? What did you decide at each time?
5. Remember the last negative emotion you experienced. Reflect on the circumstances.
6. What in your life now reproduces those early feelings and decisions?

Systems Strategies

Systems strategies are selected when the objective is to improve helpees' observation and communication skills and their relationships within and outside their family. The focus is on interpersonal processes. An individual's problem is considered only in the context of his or her relational systems, regardless of how many people in the system come for help. Thus, the focus is on interactions between individuals and the assumption is that problems and problem solving involve all members of the system.

The aim of the intervention strategies is to resolve problems by rearranging the family (or other relationship) system (such as a classroom, a work place). This requires changing the communication and relationship patterns. The systems techniques used may include **reframing** the problem from a systems perspective (such as relabeling a child's behavioral problems as helpful and positive because they bring the parents together and interrupt their fighting); teaching verbal and nonverbal communication and problem-solving skills by providing feedback, new information, and coaching; assigning direct and indirect behavioral tasks; as well as **psychoeducation** (supportively providing guidance and new information by didactic teaching and bibliotherapy). Systems therapists often choose techniques from the other major models to implement their goals. Specific techniques include the following:

1. Prescribing the symptom, such as telling a couple to fight more, not only because that will increase their caring but also because they are so good at loving that way.
2. Family sculpting, a nonverbal experiential technique in which family members position themselves or are positioned by the therapist in a tableau that reveals their perceptions and feelings about family relationships.
3. Drawing a genogram, a family map over several generations, to help understand the themes and patterns of the family's relationship style and lifestyle.
4. Circular questioning, asking nonparticipating members to speak for others or to hypothesize about the past or future in relation to the present.

The following example illustrates a systems perspective.

Ms. Marsh, age 32, was referred for help by her neighborhood health center because of continuing headaches without any organic cause. She complained of irritability and persistent arguments with her 6-year-old son. A single parent on welfare, she had not worked steadily since the birth of her child. Her son seemed to be doing well in kindergarten. At the first session, the helper learned that Ms. Marsh's 26-year-old boyfriend, a cab driver, was putting pressure on her to let him move in with her. She said she was afraid of jeopardizing her welfare benefits. The mental health worker suggested that the boyfriend come in for a joint couple session. During several couple sessions, it became clear that the headaches and irritability were symptoms of trouble in the couple system. Mr. Rich, the boyfriend, now wanted to become more involved with Ms. Marsh and her son; Ms. Marsh was fearful of this closeness due to the pain she had experienced in her brief relationship with the father of her son and in the abusive, alcoholic family from which she had run away. The couple were at an impasse. The level of their arguing was escalating. In tracking the pattern of this relationship, it emerged that every time one member of the couple wanted to move closer, the other became frightened and started arguments to create distance. Rather than continue this pursuer-distancer cycle, the couple contracted with the helper to learn to communicate their concerns more directly to each other and to engage in strategies that would allow them to get closer to each other.

When to Use Systems Strategies

The systems perspective is important in all cases, in all settings that are human systems, since no one lives in a vacuum. Whether one works with an individual, a couple, or an entire family, with a student, a student and teacher, a classroom, or an entire school, one must consider how any individual's problem contributes to the systems in which the individual exists, and how the individual simultaneously is affected by this system. It is the perspective, rather than any particular strategies, that is important.

The following exercise illustrates some of the systems principles.

Exercise 6.31   The purpose of this exercise is to facilitate your reframing a problem from a systems perspective. Try to remember a problem you had in your family growing up. What did it mean to you? To each other member of your family? Who benefited and who suffered from this problem? What were all of the ramifications of this problem? Then? Later? Now? Pick a partner in your class and share your thinking about your problem. Together, consider what you can learn about this problem from the systems perspective. What do you think about this problem now?

Exercise 6.32   Now assess where you are as a helper in relationship to the seven aspects of human personality addressed by multimodal therapy.

1. *Behavior:* Which helping behaviors do you want to learn, to increase, and/or to decrease?
2. *Affect:* What feelings do you experience most often as a helper (in class or on the job)? Which ones (such as anxiety, anger, and guilt) hinder your ability to function as a helper?
3. *Sensation:* Are there any negative sensations, such as tension, butterflies in your stomach, muscle tightness, light-headedness, blushing, rapid eye blinking, or tapping, that you experience when you are in the helper role?
4. *Imagery:* What pictures or images come to mind when you are helping? Do they bother you? Help?

5. *Cognition:* What ideas, values, opinions, and attitudes get in the way of your helping? What do you find yourself saying? You may refer to Ellis's list of irrational ideas to see which negative self-statements seem familiar.

6. *Interpersonal relationships:* Write down concerns you have about working with others, your teacher, your classmates, or your supervisor. In class exercises, do you have any problems with the classmates with whom you role play? When your teacher observes you, does that inhibit you? Are you more, or less, comfortable as helper, helpee, or observer? Discuss.

7. *Diet/drugs:* Write down any health habits or illness or effects of medication that interfere with your ability to help, such as hangovers, skipping dinner.

What do you learn about yourself as you review what you have written?

Exercise 6.33   Using Ivey's DCT model (see page 128), how would you rate the following helpee statements? Rate a statement SM (sensorimotor) if it focuses on the elements of immediate experience; CO (concrete-operational) if it searches for situational descriptions; FO (formal-operational) if it discerns patterns of thoughts, emotion, and action; and DS (dialectic-systemic) if it integrates patterns of emotion and thought into a system. After you rate these responses, write a helper response to each helpee statement that would receive the same rating you have ascribed to the helpee statement.

Helper: What part do you think you play in your marriage problems?
1. **Helpee:** She's always picking on everything and then I just shut up and go back to my paper.
2. **Helpee:** She is always on my back. She's like my mother—nag, nag, nag. I feel it in the pit of my stomach. Before I walk in the house, I feel myself tighten up.
3. **Helpee:** She really means well and I know I drive her up a wall by not always paying good enough attention. Then she gets upset and then it all goes off.
4. **Helpee:** I know I'm part of the problem. Sometimes I don't pay attention and then she gets frustrated and gets on my case. And then I withdraw and she gets more upset. It's the same kind of stuff my folks did and I swore I wouldn't be like that.

The following exercise serves as a review of the criteria on which to base your choice of approach. Remember, the more strategies you can learn to use effectively, the wider the variety of problem situations you can help with.

Exercise 6.34   Based on your understanding of this chapter, circle the letter of the strategy or strategies you would consider most appropriate for each of the following cases. Then, in small groups, discuss the reasons for your selections. Remember, there are no hard-and-fast rules, except for some very specific cases, and even then more than one strategy can be applied. Possible answers are discussed at the end of the chapter.

1. Lisa, age 18, is deathly afraid of flying. She is unable to identify the source of this fear, but she is concerned because her fiancé has accepted an out-of-town job, and she is expected to fly in to see him once a month for the next year.
   a. client-centered therapy
   b. rational-emotive therapy
   c. systematic desensitization
   d. reality therapy
   e. Gestalt therapy
   f. transactional analysis

2. Ms. Carleton is feeling anxious because her supervisor has unexpectedly left and the new supervisor hasn't arrived yet. She was very close to her old supervisor and doesn't know anything about the new one. Her work is suffering and she is scared.
   a. client-centered therapy
   b. rational-emotive therapy
   c. assertiveness training
   d. reality therapy
   e. Gestalt therapy
   f. transactional analysis

3. Mr. Wright, age 45, married with three children, has just been told he'll have to take a 20 percent salary cut in order to keep his job. The firm for which he works has been having difficulty for several months. As Mr. Wright's living costs have risen considerably due to inflation, he does not think he can manage on this reduced salary.
   a. client-centered therapy
   b. rational-emotive therapy
   c. decision-making therapy
   d. reality therapy
   e. Gestalt therapy
   f. systematic desensitization

4. Susie, age 13, is causing her mother great distress because of her low marks and sassy conduct at school and at home. Her mother has come for help because she is "at the end of her rope" and doesn't know what to do.
   a. client-centered therapy
   b. rational-emotive therapy
   c. contract therapy (behavioral)
   d. reality therapy
   e. Gestalt therapy
   f. transactional analysis

5. Phyllis, a 34-year-old divorcée, is unable to pay child-care costs for her son on her salary. However, if she quits work, she won't have any income, because her ex-husband is out of work and can't help out at all. She doesn't know what to do and she is very depressed.

    a. client-centered therapy
    b. rational-emotive therapy
    c. assertiveness training
    d. reality therapy
    e. Gestalt therapy
    f. transactional analysis

6. Sandra's father has told her that she must quit college and get a job. He does not believe that women should go to college and is worried that he won't be able to provide tuition for Sandra's younger brother, who is a junior in high school. Sandra has been an average student, is in her sophomore year at school, and, while she wants to stay in school, sees no way that she can disobey her dad.
    a. client-centered therapy
    b. rational-emotive therapy
    c. assertiveness training
    d. reality therapy
    e. Gestalt therapy
    f. transactional analysis

7. Mr. Brown feels that his boss is constantly criticizing him in front of others in the office. He works in a small office, can't really get out of sight of his boss, and has become increasingly nervous and distraught.
    a. client-centered therapy
    b. rational-emotive therapy
    c. assertiveness training
    d. Gestalt therapy
    e. reality therapy
    f. transactional analysis

8. Martha, age 15, is very angry with her mother. She says her mother is unfair, too punitive, and not at all understanding. Martha's marks in school have dropped considerably this year and she is much moodier than previously.
    a. psychoanalysis
    b. client-centered therapy
    c. rational-emotive therapy
    d. assertiveness training
    e. reality therapy
    f. Gestalt therapy

9. Tom, age 14, comes for help because he is scared. Several of his friends were involved in a house break-in over the weekend, and although he didn't go into the house, he was outside waiting for his friends to come out. Everybody in town is talking about the vandalism this gang did in this house, and Tom is afraid he'll be implicated sooner or later.
    a. psychoanalysis
    b. reality therapy
    c. rational-emotive therapy

    d. client-centered therapy

    e. Gestalt therapy

    f. systematic desensitization

10. Mary Lou, age 23, is distressed because another love affair has just ended. She feels very sorry for herself and resents that this always happens to her. "Every time I get close to or begin to get close to someone, he leaves me."

    a. psychoanalysis

    b. rational-emotive therapy

    c. modeling

    d. reality therapy

    e. Gestalt therapy

    f. transactional analysis

11. Larry, age 22, is very angry and bitter because he did not get into graduate school. He feels he had to settle for second best in college, too. Larry says, "I never get what I want. I always end up with second best."

    a. psychoanalysis

    b. rational-emotive therapy

    c. reality therapy

    d. client-centered therapy

    e. transactional analysis

12. Marianne, age 32, is severely depressed and has not eaten or slept more than a few hours in days. She has a long history of illnesses and a high incidence of absenteeism from work. She doesn't know what's bothering her and just wants to be left alone.

    a. psychoanalysis

    b. client-centered therapy

    c. reality therapy

    d. systematic desensitization

    e. Gestalt therapy

    f. transactional analysis

13. Max, age 18, has been accepted by the four colleges to which he applied. Although he is overjoyed, he is baffled by the need to choose a college, and has become increasingly nervous and irritable as the deadline approaches.

    a. reality therapy

    b. decision-making therapy

    c. client-centered therapy

    d. rational-emotive therapy

    e. assertiveness training

    f. Gestalt therapy

14. Ms. Wolfe is the divorced mother of three young children. She's come for help because she can't manage her children and is always losing her temper and slapping them. She feels that 100 percent of her time and energy is devoted to motherhood and that she has no time for herself.

    a. contracting

  b. rational-emotive therapy

  c. psychoanalysis

  d. systematic desensitization

  e. transactional analysis

  f. client-centered therapy

15. Mr. Winters has recently been released from a mental hospital. He is having trouble finding work. He is very nervous and always lets others get ahead of him in employment lines.

  a. rational-emotive therapy

  b. assertiveness training

  c. transactional analysis

  d. modeling

  e. client-centered therapy

  f. Gestalt therapy

Exercise 6.35 Now go back over these 15 cases and see which of the modalities from the BASIC ID model (page 128) you would emphasize as priorities in each case.

☐ Summary

This chapter's brief overview of strategies suggests the skills and some of the criteria necessary for selection of strategies. We presented examples and exercises to provide an initial exposure to representative strategies emanating from the helping theories discussed in Chapter 5. Helpers modify and adjust various combinations of effective strategies according to their own personalities and preferences and the needs of the helpee.

  The strategies discussed in this chapter range from the commonsensical (such as decision making) to the complex and powerful. *All* strategies require training, supervision, and experience to be used effectively. Again, it cannot be stressed enough that there is no one strategy to fit any particular problem. For example, in Exercise 6.34 some strategies appeared to be more effective than others for each of the situations presented, but a skillful helper could have effectively adopted many of the possible strategies for any one problem. The best guidelines for choosing strategies are to learn from your own and others' experiences and to risk whatever seems right to you at the particular moment.

  The model upon which the categorization of strategies in this chapter is based is the affective-cognitive-behavioral continuum. The helper identifies which domain is the primary context for the problem and which holds most promise for resolving it. Everyone's ultimate goal is to function effectively in all three domains, and it is usually clear which domain is weak and in need of strengthening in any given individual.

☐ Exercise Answers

Exercise 6.18   Numbers 2 and 4, the second part of 6, and numbers 11, 13, and 15 are behavioral statements. The others express subjective conclusions, attitudes, and evaluations.

Exercise 6.24   1. a, 2. c, 3. b, 4. c, 5. b, 6. c, 7. a, 8. b, 9. c, 10. b

Exercise 6.33   The first helpee statement is CO in that the helpee is able to describe concretely his experience but not experience his own feelings. There is no reflection about his part in the marital conflict and no ability to recognize any patterns. The second helpee statement is SM in that he is lost in his feelings and has no real understanding of what is going on. The third statement is FO in that he is able to identify some of the patterns in his marriage and present them in a more abstract manner as opposed to concrete description of a specific incident. The fourth helpee statement is DS in that the helpee is able to be more analytical in recognizing the patterns in his marriage and how they resemble those in his parents' marriage. By learning to assess the cognitive-developmental level of helpee statements and of helper responses, helpers can learn to match their helping style to the developmental needs of the helpee.

Exercise 6.34   Possible answers are as follows:

1. Systematic desensitization has been proven to be effective with phobias. One can use client-centered therapy at the same time.
2. I would opt for client-centered therapy here, involving the client in a relationship that would reduce her anxiety.
3. Some decision-making and/or reality therapy would be helpful here, examining all the options and values involved so that Mr. Wright can accept responsibility for taking the cut or looking for a new job.
4. Teaching the mother how to use behavioral contracts may clear up some of the hassling that is occurring at home. We don't know enough to know what the problem is or whose problem it is, but if Susie becomes the client, reality therapy may be helpful, and if the mother remains the client, TA can be helpful.
5. Except for assertiveness training, any of the strategies could be appropriate in this case. Client-centered therapy could help her feel better about herself, rational-emotive therapy could get her in touch with her irrational ideas, reality therapy could help her with choosing responsible behavior, and Gestalt and TA could enhance her emotional and intellectual awareness.
6. Assertiveness training could help Sandra deal with her father; rational-emotive therapy could help her examine her belief system about families,

authority, and education; and reality therapy could help her make responsible choices.

7. Assertiveness training along with some rational-emotive therapy could be effective here.
8. In cases such as this, I have found a combination of client-centered therapy and Gestalt therapy to increase insight and self-concept.
9. Reality therapy and client-centered therapy can be helpful here: the former to help Tom evaluate his own behavior and the latter to help him feel good about himself.
10. Psychoanalysis or transactional analysis could be helpful here to enable Mary Lou to gain some insight into patterns of relating.
11. Again, psychoanalysis or transactional analysis could be helpful here. I'd opt for the TA, because it sounds as if Larry has made an early decision never to be successful.
12. Psychoanalysis could be helpful here if this has been going on for a long time. Reality therapy could be effective, too, in forcing Marianne to assume some responsibility for her behaviors.
13. Decision-making therapy is the obvious choice here.
14. Some rational-emotive therapy might help here. It sounds as if Ms. Wolfe has some "crazy" ideas about motherhood. Some contracting principles could help her with child management, and some TA could give her insight into where she's coming from.
15. Some client-centered therapy could help him gain self-confidence. Assertiveness training would also be useful.

These answers are only suggestions and are open to discussion.

☐ References and Further Reading

General

Bruce, P. (1984). Continuum of counseling goals: A framework for differentiating counseling strategies. *Personnel and Guidance Journal 62,* 259–264.

Corey, G. (1991). *Case approach to counseling and psychotherapy.* (3rd ed.). Pacific Grove, CA: Brooks/Cole.

Corsini, R. J., & Wedding, D. (Eds.). (1989). *Current psychotherapies* (4th ed.). Itasca, IL: Peacock.

Kutash, I. L., & Wolf, A. (Eds.). (1986). *Psychotherapist's casebook.* San Francisco: Jossey-Bass.

Okun, B. F. (1990). *Seeking connections in psychotherapy.* San Francisco: Jossey-Bass.

Zeig, J. K. (Ed.). (1989). *The evolution of psychotherapy.* New York: Brunner/Mazel.

Affective-Cognitive Strategies

Blocher, D. (1966). *Development counseling.* New York: Ronald.

Bordin, E. (1968). *Psychological counseling.* New York: Appleton-Century-Crofts.

Cashdan, S. (1988). *Object relations therapy.* New York: Norton.

Freud, S. (1943). *A general introduction to psychoanalysis.* Garden City, NY: Doubleday.

Freud, S. (1949). *An outline of psychoanalysis.* New York: Norton.

Holzman, P. (1970). *Psychoanalysis and psychopathology.* New York: McGraw-Hill.

St. Clair, M. (1986). *Object relations and self psychology: An introduction.* Pacific Grove, CA: Brooks/Cole.

Watkins, C. H., Jr. (1983). Transference phenomena in the counseling situation. *Personnel and Guidance Journal, 62,* 206–210.

Affective Strategies

Client-Centered

Carkhuff, R., & Truax, C. (1967). *Toward counseling and psychotherapy: Training and practice.* Chicago: Aldine.

Combs, A. W. (1989). *A theory of therapy: Guidelines for counseling practice.* Newbury Park, CA: Sage.

Levant, R. F., & Shlien, J. M. (Eds.). (1984). *Client-centered therapy and the person-centered approach: New directions in theory, research, and practice.* New York: Praeger.

Patterson, C. (1974). *Relationship counseling and psychotherapy.* New York: Harper & Row.

Rogers, C. (1951). *Client-centered therapy.* Boston: Houghton Mifflin.

Rogers, C. (Ed.). (1967). *The therapeutic relationship and its impact.* Madison: University of Wisconsin Press.

Gestalt

Fagan, J., & Shepherd, I. (Eds.). (1970). *Gestalt therapy now.* Palo Alto, CA: Science and Behavior Books.

Levitsky, A., & Perls, F. (1970). The rules and games of Gestalt therapy. In Fagan & Shepherd (Eds.), *Gestalt therapy now* (pp. 140–150). Palo Alto, CA: Science and Behavior Books.

Passons, W. (1975). *Gestalt approaches in counseling.* New York: Holt, Rinehart & Winston.

Perls, F. (1973). *The Gestalt approach and eye witness to therapy.* Palo Alto, CA: Science and Behavior Books.

Polster, I., & Polster, M. (1973). *Gestalt therapy integrated.* New York: Brunner/Mazel.

Zinker, J. (1978). *Creative process in Gestalt therapy.* New York: Random House.

Cognitive and Cognitive-Behavioral Strategies

Beck, A. T. (1976). *Cognitive therapy and emotional disorders.* New York: New American Library.

Ellis, A. (1962). *Reason and emotion in psychotherapy.* New York: Lyle Stuart.

Ellis, A., & Dryden, W. (1987). *The practice of rational-emotive therapy.* Secaucus, NJ: Lyle Stuart.

Ellis, A., & Grieger, R. (1986). *Handbook of rational-emotive therapy: Vol. 2.* New York: Springer.

Ellis, A., & Whiteley, J. (Eds.). (1979). *Theoretical and empirical foundations of rational-emotive therapy.* Pacific Grove, CA: Brooks/Cole.

Glasser, N. (Ed.). (1989). *Control theory in the practice of reality therapy: Case studies.* New York: Harper & Row.

Glasser, W. (1986). *The control theory-reality therapy workbook.* Canoga Park, CA: Institute for Reality Therapy.

Glasser, W. (1990). *The quality school.* New York: Harper & Row.

Maultsby, M. C. (1984). *Rational behavior therapy.* Englewood Cliffs, NJ: Prentice-Hall.

Meichenbaum, D. (1977). *Cognitive behavior modification: An integrative approach.* New York: Plenum.

Wubbolding, R. E. (1988). *Using reality therapy.* New York: Harper & Row.

Behavioral Strategies

Alberti, R., & Emmons, M. (1986). *Your perfect right: A guide to assertive action.* (5th ed.). San Luis Obispo, CA: Impact.

Goldfried, M. R., & Davison, G. C. (1976). *Clinical behavior therapy.* New York: Holt, Rinehart & Winston.

Homme, L. (1970). *Use of contingency contracting in the classroom.* Champaign, IL: Research Press.

Kanfer, F. H., & Goldstein, A. P. (Eds.). (1986). *Helping people change: A textbook of methods.* New York: Pergamon Press.

Kazdin, A. E. (1980). *Behavior modification in applied settings.* (rev. ed.). Homewood, IL: Dorsey Press.

Krumboltz, J., & Thoresen, C. (Eds.). (1976). *Counseling methods.* New York: Holt, Rinehart & Winston.

Sulzer-Azaroff, B., & Mayer, G. (1977). *Applying behavior analysis procedures with children and youth.* New York: Holt, Rinehart & Winston.

Ullman, L., & Krasner, L. (1965). *Case studies in behavior modification.* New York: Holt, Rinehart & Winston.

Watson, D. L., & Tharp, R. G. (1989). *Self-directed behavior: Self-modification for personal adjustment* (5th ed.). Pacific Grove, CA: Brooks/Cole.

Wolpe, J. (1969). *The practice of behavior therapy.* New York: Pergamon Press.

Strategies That Cut across Domains

Transactional Analysis

Berne, E. (1964). *Games people play.* New York: Grove Press.

Berne, E. (1972). *What do you say after you say hello?* New York: Grove Press.

Freed, A. (1972). *TA for kids.* Sacramento, CA: Jalmar.

Freed, A. (1972) *TA for tots.* Sacramento, CA: Jalmar.

Goulding, R., & Goulding, M. (1978). *The power is in the patient: A TA/Gestalt approach to psychotherapy.* San Francisco: TA Press.

James, M., & Jongeward, D. (1971). *Born to win.* Reading, MA: Addison-Wesley.

James, M., & Jongeward, D. (1973). *Games for winning.* Reading, MA: Addison-Wesley.

Shepard, M., & Lee, M. (1970). *Games analysts play.* New York: Berkley-Medalion.

Woollams, S., & Brown, M. (1979). *TA: The total handbook of transactional analysis.* Englewood Cliffs, NJ: Prentice-Hall.

Eclectic and Multimodal Strategies

Ivey, A. E. (1991). *Developmental strategies for helpers: Individual, family, and network interventions.* Pacific Grove, CA: Brooks/Cole.

Lazarus, A. (1976). *Multimodal behavior therapy.* New York: Springer-Verlag.

Lazarus, A. (1989). *The practice of multimodal therapy.* Baltimore, MD: Johns Hopkins University Press.

Systems Strategies

Haley, J. (1976). *Problem-solving therapy.* San Francisco: Jossey-Bass.

Madanes, C. (1984). *Strategic family therapy.* San Francisco: Jossey-Bass.

Minuchin, S., & Fishman, H. C. (1981). *Techniques of family therapy.* Cambridge, MA: Harvard University Press.

Satir, V. (1967). *Conjoint family therapy* (2nd ed.). Palo Alto, CA: Science and Behavior Books.

Wachtel, E. F., & Wachtel, P. L. (1986). *Family dynamics in individual therapy: A guide to clinical strategies.* New York: Guilford Press.

SEVEN

Stage 2:
Applying Strategies

The boundary line between the first (relationship) and the second (strategy) stages of the helping relationship is not distinct. You will know when you are at the point of crossing the boundary when you are able to focus on identifiable problems, when you find yourself thinking "O.K., now that we know what the trouble is, what can we do about it?" Remember that you do not want to rush into the strategy stage until you have clearly ascertained and clarified the helpee's problem and until you have agreement from the helpee that he or she wants to do something about it.

In this chapter we will examine the six steps in applying strategies during the second, strategy stage of the helping relationship. These six steps are (1) mutual acceptance of defined goals and objectives, (2) planning of strategies, (3) use of strategies, (4) evaluation of strategies, (5) termination, and (6) follow-up. At the end of the chapter, we'll examine some case studies that will take you through the steps of both the relationship stage and the strategy stage.

☐ Step 1: Mutual Acceptance of Defined Goals and Objectives

The helping relationship must focus on areas of concern to the helpee, not on concerns that the helper thinks the helpee ought to work on. This necessity may become a serious issue in some organizational settings where a conflict may exist between organizational policies, which you as the helper may represent, and the needs of the helpee. For example, if you are working in a correctional institution and an inmate who is causing some disruption and is seen as a "troublemaker" who must be changed is sent to you, you may find that you and this inmate have different objectives. Yours may be, of institutional necessity, to help the inmate conform to the system, whereas the inmate's objectives may be to disrupt and get whatever attention results from the disruption. No one can tell you how to handle this kind of conflict, but you really have to think it out and come to a decision with which you are comfortable.

It is important to have a theoretical rationale for your understanding of the client's problem and definition of goals and objectives. This theoretical framework will help lead you to a specific choice of strategies. Also remember that in order to be of help you must have a strong helping relationship, and that building this relationship involves identifying goals and objectives that are accepted by all involved parties. Strategies will not be effective if a strong helping relationship does not exist.

The following examples illustrate negative outcomes when this first step is not attended to.

Mary and Tom entered counseling with an accredited marriage counselor for intensive sex therapy. Their objective was to improve their sexual relationship. The counselor felt that he needed to explore the dynamics of their marital relationship, because he believed the underlying problems in the marital relationship were maintaining the sexual dysfunction. When Mary and Tom achieved success through some behavioral sex therapy, they wished to terminate the therapy. They became upset when the counselor suggested they remain in treatment for more intensive marriage counseling. The point is that the helper and the helpees had not clarified the nature and extent of the problem and treatment. Thus, what began as an effective helping relationship ended with confusion and misunderstanding on both sides.

Lisa, age 25, was admitted to a state rehabilitation counseling program for job training after her release from a state hospital psychiatric unit. Her understanding was that this program consisted of coursework and job counseling as well as placement advice. When the six-week group human relations course was completed, Lisa expected to receive some resumé and job seeking skill counseling. The counselor told her to look at the ads in the newspaper every day and that he would check in with her by phone once a week. After several weeks of no contact from the counselor and no job possibilities, Lisa's symptoms of suicidal depression began to reappear. Investigation by the welfare worker indicated that the printed information about this program, the verbal promises of the counselor, and the actual circumstances were incongruent. Again, the lack of clearly specified goals and objectives resulted in pain and illness rather than growth and health.

John went to a vocational counselor for tests, to help him make a career decision. He was very disturbed when the counselor tried to get him to talk about his personal feelings and lifestyle instead of presenting him with factual results of his interest and aptitude tests. After the third session, he quit and decided to go elsewhere. This is another example in which the nature of the problem and approach was not fully clarified.

Mrs. Escadero agreed to have Christina, age 8, work with a volunteer aide for individualized help. Mrs. Escadero thought that Christina would receive some academic skills help in reading and arithmetic. The aide was told by the classroom teacher that her objective was to "develop a relationship with Christina to improve her self-concept." The aide played games with Christina and took her for walks. Christina's social skills improved and she seemed happier in her class, but her mother became angry when she found out that academic tutoring was not occurring and insisted that the relationship terminate. This case is still another example of what may happen when objectives of a helping relationship are not clarified.

Exercise 7.1   Discuss in small groups how you could handle the preceding four cases in a different manner. Act out different approaches and share your feelings and reactions. How can you be sure that the helpee really accepts goals and objectives and is not just trying to please you?

Here is another example of the importance of the first step of the strategy stage.
Lauren was referred to a counselor in May of her senior year in high school. The presenting problems were promiscuity and possible pregnancy (however, it turned out that Lauren was not pregnant). In the first three sessions, the counselor discovered that Lauren had a very low self-concept, received little, if any, positive feelings from her parents (who were divorced), had no close girlfriends despite her popularity, and was, in fact, quite the opposite of her outward image of a beautiful, talented, bright, "model" teenager. Lauren has just expressed some concerns about going away to college in the fall.

Counselor: You're wondering if you're going to be able to find the kind of friends and relationships you want there without getting into trouble.
Lauren: I'm not sure I'll be able to handle difficult situations when they come up. I always get messed up.
Counselor: You really don't think much of yourself, do you?
Lauren: *(laughs nervously)* I know . . . it's just that sometimes I try to think so much, I get awfully confused.
Counselor: Well, we have eight weeks left before you leave, and we've agreed to meet weekly. How about if we focus on talking about your concerns and situations as they come up, and see if we can help you to feel better about yourself before you go away.
Lauren: I'm already feeling better since I've been coming in here. I do want to make it on my own next year.

In this case, the counselor and Lauren decide that the major goal will be to increase Lauren's self-confidence and to afford her the opportunity to receive empathy and support from the counselor.

☐ Step 2: Planning of Strategies

The second step of the strategy stage is similar to the decision-making process in that in both instances we think of all the possible options and then select the most effective strategy or combination of strategies. We try to ascertain which strategy is best for *this* individual at *this* time, given our understanding of the client, his or her problem, our theoretical orientation, and the treatment context. The client's participation in this decision making increases the possibilities for successful outcomes. We carefully explain to the helpee the approaches we are considering, their rationale, possible consequences, time and activities involved, and any other pertinent information. If the helpee seems to be resistant to some strategies, we discuss the resistance and then decide together whether to hold off or tentatively try them out. Some strategies may seem threatening at first but become less so as the helping relationship develops. For example, I often use Gestalt techniques and, unless the relationship is well established and the client trusts me and my professional ability, a common response to the suggestion that we engage in a particular Gestalt dialogue or role-play is "That's silly; I can't do that." Sometimes helpees get scared away because they do not understand what is going on and strategies have not been adequately explained. If the helper explains different techniques in a tentative way, the helpee can refuse without losing dignity or self-respect.

Thus, timing and manner of introducing alternative approaches are important factors in the planning of strategies, as illustrated in the following example.

> **Helper:** We seem to agree that one of the immediate goals you want to work toward is to be able to say no to people, to see if we can find some other ways for you to feel good about yourself so you won't always have to buy friendship.
>
> **Client:** Yeah. My way certainly hasn't worked either.
>
> **Helper:** Evelyn, I wonder if we can do some role playing and practice what you might say to Benny the next time he calls you.
>
> **Client:** Sure. What do you mean?
>
> **Helper:** Well, I'll be Benny, and I've called you to have you come up for the weekend. Now what do you really want to say to me?

Here the helper has left the helpee an outlet. The relationship is such that the helpee knows she will not be rejected if she does not go along, but because she believes that the helper knows what she is doing, she is willing to go along and try something new.

Another example of strategy planning can be drawn from the case of Lauren.

Counselor: Lauren, in addition to talking about you and what's happening in your life, I'd like to ask you to keep a diary. If each night you'll write down all the things that happened that day to make you feel good about yourself, it will be helpful.

Lauren: I don't know if there'll be any, but I'm willing to try.

Counselor: We can go over your list when you come in. It might help us both to find out what pleases you. To get started, I'd like you to draw up a list of all the things you like about yourself. You don't have to show it to anyone, and you can bring it in next week.

Lauren: O.K.

The purpose of this "homework assignment" is to help Lauren focus on positive aspects of herself. It also provides some continuity between sessions.

Exercise 7.2 How would you plan strategies for the helpees in the preceding two examples? Role play the scenes, and then create your own cases that you can act out. Discuss in small groups the criteria and reasons for your selections. Remember that there is not just one technique for any one problem. There are times when helpers try something that both parties agree to and find it doesn't work very well. However, if the relationship is effective, no one need lose face, and the strategy-planning step can be repeated.

☐ Step 3: Use of Strategies

A particular strategy or combination of strategies may be used for short-term or long-term goals. The helper may spontaneously decide to change strategies, and this is fine, as long as the helpee knows what is going on and agrees to the change. Often, the helper's intuition sparks appropriate selection of strategies midway through this step. For example, several years ago I was using systematic desensitization with a college physical education student who had a phobic reaction to balloons and who found that she needed to be able to tolerate being in the same room with balloons during some school affairs. She was an excellent candidate for systematic desensitization and responded beautifully. One day after a successful session, while she was still relaxed, I asked her to engage in some Gestalt dialogues with balloons. She was equally responsive to this technique, and we both learned some important information about her aversion to balloons. I also learned that relaxation facilitates Gestalt dialoguing.

As you gain experience as a helper, your self-confidence in the use of strategies will increase and you will feel more comfortable trying out new techniques. However, do not be afraid to let the helpee know if you are trying something you have not used before. You can say something like "I have a new idea that we can try, but I'm not sure how it will work. Let's try it out and see if it works." Perhaps you can arrange for some supervision or consultation from someone more experienced with a particular strategy. Pretending to be an expert when you're not can backfire, and even if it doesn't, it is ethically questionable. However, if you restrict yourself to the familiar and comfortable, you won't be able to extend your capabilities. In any case, for both your benefit and the helpee's, any significant move into new strategies should be accomplished carefully and under the guidance or supervision of someone who is qualified in that particular area.

The amount of time available for a helping relationship affects the use of strategies. By this I mean the amount of time both for each session and for the entire relationship. It is impossible to say how much of the entire helping relationship is devoted to using strategies—it may be a little; it may be a lot. It depends on many other variables, the most important being the nature of the problem and your setting. Some strategies can be used in a shorter time span than others; behavioral strategies are relatively short-term, whereas psychoanalytic and client-centered strategies can last indefinitely. It seems that those strategies that lean toward the behavioral domain are more specifically evaluated and easier to limit to a particular time frame, whereas those strategies that lean toward the affective domain take a longer time for application.

Strategies facilitate learning. Coupled with a strong helping relationship, they can expedite the helpee's emotional, cognitive, or behavioral changes.

Exercise 7.3   The purpose of this exercise is to give you some limited experience in planning and using strategies. In triads, rotate the roles of helper, helpee, and observer. You can choose your settings and circumstances, but assume that both helper and helpee have agreed to one of the following goals for the helping relationship. After selecting an objective, role play it as far as you can. What strategies would you want to use? How would you want to use them?

Here are the goals:

1. To learn how to make friends of the opposite sex
2. To get homework assignments in on time and regularly
3. To learn to get along better with the boss
4. To decide whether or not to have a baby
5. To try to control anger when things don't go your way

Assume that you already have an effective relationship. Discuss your choice and use of strategies in small groups.

☐ Step 4: Evaluation of Strategies

Evaluation (judging the effectiveness) of strategies is ongoing from the beginning of their use. If at any time both helper and helpee feel that the strategies are not effective, it is necessary to review and reassess the situation. However, it sometimes takes a while for something to work, and a sensitive helper knows when to apply evaluation criteria.

Evaluation of counseling's effectiveness has always been difficult because counseling strategies (except for behavioral strategies) do not always produce observable changes. Actual behavior changes provide the best criteria for evaluating strategies. These behavior changes can be reflections of emotional and cognitive changes. If we do not look for behavior changes, we may be fooled by some people who say they feel good because they want to please us. However, we can always tell if something has been effective by observing behavior.

A behavior change does not occur just within the helping relationship; it also generalizes to other situations in the helpee's life. The helpee may report this change or the helper may hear about it from significant others, such as a supervisor, parent, spouse, or teacher. Here again, the effectiveness of the helping relationship is important. The helper and helpee can objectively seek criteria for evaluating helping and can both validate these criteria, rather than take effectiveness for granted. Time is important because it is realistic to expect some ups and downs during the course of a helping relationship, and new behaviors, attitudes, and thoughts take time to become stable.

An evaluation may cause the helper to decide to use another strategy, work on some other objectives and goals, or terminate the helping relationship. The helpee participates in the selection of one of those outcomes.

The following example illustrates the strategy-evaluation step.

Dr. J. is a 52-year-old dentist who came for help when his 22-year-old son dropped out and joined a religious cult. Terribly distressed, Dr. J. didn't understand what went wrong and how this could have happened to his family. The counselor has been using rational-emotive techniques and assigning homework to Dr. J. to dispute his irrational ideas. This excerpt is from the ninth session.

Helper: I'm wondering how things are going at home for you now.

Client: They're beginning to change. Every time I start to think or say that my wife or one of the kids *should* be or do something, I catch myself and shudder at that awful word "should."

Helper: And what happens?

Client: I don't get so angry anymore. I keep telling myself that everyone doesn't have to be perfect all the time and neither do I. Also, things don't always have to go my way. I'm beginning to believe it.

Helper: *(laughs)* Good. You're learning.

Client: Well, my wife says things are much pleasanter, and we don't have so much quarreling at the dinner table.

In this case, the helper is evaluating the effectiveness of some strategies and, in addition to learning whether the strategy is working, gaining some insights into future directions of this helping relationship.

Exercise 7.4   The purpose of this exercise is to familiarize you with evaluative criteria. For the following objectives of helping relationships, pick out what you think would be appropriate evaluative criteria. Answers are at the end of the chapter. Discuss your answers and reasons for them in small groups.

1. Objective: Learning to get along better with co-workers
    Criteria: a. Fewer complaints from supervisor
            b. Positive report from helpee
            c. Your observation of friendlier relationships in department
            d. Your observation of helpee in cafeteria engaged in friendly relations with co-workers
2. Objective: Feeling less depressed
    Criteria: a. Client improves appearance
            b. Client reports feeling better
            c. Client reports more involvement in activities
            d. Client doesn't cry as much
3. Objective: Increasing self-understanding
    Criteria: a. Client reports feeling better
            b. Client wants to continue sessions
            c. Client talks in sessions with more assurance and confidence
            d. Client reports better interpersonal relationships
4. Objective: Learning to make friends of opposite sex
    Criteria: a. Client talks about going to singles' bar
            b. Client reports making date for tonight
            c. Client asks if your receptionist is single
            d. Client says he has struck up conversation with a girl in his class
5. Objective: Improving marital happiness
    Criteria: a. Client says spouse doesn't nag so much anymore
            b. Clients report they spend more time together
            c. Clients say during counseling sessions that their communication is good outside of sessions
            d. Clients report they try but never complete homework exercises

☐ Step 5: Termination

Termination is an important part of the helping relationship; the concept applies to individual sessions as well as to the overall helping relationship. If termination is not recognized and handled appropriately, helpees may end up with more distress and unresolved issues than when they entered the helping relationship.

Termination of Session

It has been said that one can tell how a counselor will handle the termination process by how she or he handles the end of each session. Several minutes before the end of each session, it is important to communicate nonverbally and/or verbally to the helpee that the allotted time is almost up. This allows for summarization of what has occurred during the session, to be sure that both parties are clearly in accord, and for assigning in-between session tasks or planning for future sessions. In sessions that have been particularly "heavy" or emotional, both helper and helpee need time to regain composure so as to comfortably make the transition back to the outside world.

Termination of the Helping Relationship

Evaluation of the results or progress of the application of strategies may lead to termination of the helping relationship. Termination can occur in one of four ways.

1. Both helper and helpee feel that all objectives have been reached. This is a positive termination, albeit one involving sadness because of the loss of a meaningful relationship. Sad feelings usually occur in both helper and helpee, and both may experience some separation anxiety about the impending loss of a significant other. These feelings can be explored and shared so that both can leave each other feeling the sense of growth and satisfaction that comes from the achievement of objectives. It is important for both parties to know when the last meeting will occur and to leave enough time to discuss their feelings.

2. The helper initiates termination before the objectives have been achieved. This happens often in schools when a term ends or in settings where the personnel changes frequently and rapidly. In these cases it is essential that the helper inform the helpee of the impending termination and that they both fully explore the feelings aroused by this necessity: the helpee often feels angry and rejected, and the helper often feels guilty and uncomfortable. Ideally, the helper effects a referral before termination occurs. If the helper has had an effective relationship with the helpee, it will be that much easier for the helpee to work effectively with someone else.

3. Termination is necessitated and determined by a third party to the helping process. For example, a health care plan allows one or two sessions for evaluation. The helper and the helpee agree on the nature, goals, and objectives of the helping relationship and the health plan administrator informs the helper and the helpee how many sessions will be allowed. In this type of situation, the number of sessions may be known in advance of the strategy phase so that adjustments may be made in the selection and implementation of strategies or anytime during the strategy phase. Though the termination may be in the best economic interests of the health care plan, it

is not necessarily in the best interests of the helpee or in accordance with the helper's professional judgment.

4. The helpee terminates the relationship prematurely. In this case, the helpee may be escaping from a threatening situation, leaving the helper feeling useless and inadequate. If premature termination seems likely, the helper must try to determine whose problem this premature termination is; it might be the helpee's problem if he or she refuses to work anymore and chooses avoidance over approach. On the other hand, it might be the helper's problem in that he or she has been unable to develop an effective helping relationship or has selected an inappropriate strategy.

In any of these kinds of termination, helpers can provide positive support to the helpee and communicate interest in being of help should the need for help arise in the future.

For the first two types of termination, Ward (1984) has conceptualized four steps that can facilitate and strengthen outcomes of the helping relationship.

Evaluation of goal attainment is a mutually shared process whereby helper and helpee specifically check to see whether the presenting problems or symptoms have been reduced or eliminated. The helpee is encouraged to list specific behavioral changes made during the helping process and to discuss his or her thoughts and feelings about those changes. Helpees may be asked to refer to previous assignments, such as a written journal or audio or video tapes, or to check out changes with significant others. The next part of this assessment process can be to talk about how resolution of the presenting problems has affected the helpee's self-esteem, relationships with others, work or school functioning, and ability to plan the next steps in life. This step may take several sessions, particularly for helpees working on growing-up, individuation-separation, and dependency-independence issues.

Closure of relationship issues requires talking about the helpee's feelings about the helper and the helping relationship. These feelings probably can be related to other relationships, separations, and losses in the helpee's life experience and may arouse powerful emotions. When appropriate, the helper can disclose his or her feelings about the helpee and impending termination.

This is a step that requires much discussion and contributes to achieving the ability to express and experience an appropriate and meaningful good-bye to a significant person.

Preparation for self-reliance and transfer of learning includes specific planning for the future. How is the helpee going to take his or her new learning to the outside world? What kinds of supportive relationships can the helpee develop and maintain? One client has told me that she has internal dialogues with me whenever she gets into a bind. Other clients contract with themselves or others. I often ask clients to imagine future problems and deal with them through imagery in a session.

The final session is the culmination of the preparation and often includes lighter, more social discussion. Often, I give the client a memento, such as a poster, that is significant as a reminder of the work we've done together. We specifically plan for follow-up evaluation and discuss whether or not there is a possibility of follow-up telephone or session contact.

Some helpers extend terminations over several weeks. For example, they may lengthen the time between sessions by reducing weekly sessions to biweekly sessions, then to monthly sessions, then to one session every three months. Extending termination has the effect of reducing anxiety and providing follow-up and evaluation while maintaining necessary support. However, helpers must be aware of the possibility that the helpee's dependency needs may be prolonging the helping relationship. Helpers may need to set some limits in order to avoid reinforcing those dependency needs.

The same conditions necessary for the success of the initial contact in a helping relationship (encouragement, warmth, focusing on the helpee's needs) are necessary for the final contact. It is all right for you to share your true feelings with helpees and let them know that you will miss them, too.

The following is an excerpt from the end of the final session with Lauren.

Counselor: You've come a long way this summer.

Lauren: I feel so good about myself, and I'm looking forward to college. Did I tell you I got the name of my roommate?

Counselor: No, but that does make it more real. You're looking forward to making new friends and a new life.

Lauren: I think everything's going to be fine. If I need to, can I come see you?

Counselor: Of course, you know how to reach me. Listen, you've worked through some tough situations this summer. We both know you can do it.

Lauren: Yes, that's true. I'm going to miss you.

Counselor: I'm going to miss you, too. Let me know how you're doing, O.K.?

Lauren: Sure thing.

The counselor leaves the door open in case Lauren wants to come back at any time, and feels free to express sadness at the end of the relationship.

If the client abruptly ends the relationship by not showing up for an appointment and not contacting the counselor, the counselor may attempt contact by phone or letter to see if the client was testing his or her concern and interest or needed reassurance for some other reason; it may be possible to schedule at least one more session to work through unresolved issues, effect a referral, or at the very least gain some understanding of the reasons for termination.

Exercise 7.5   Pick as a partner someone you feel particularly close to, someone you have shared with. Find a place where you can spend some time together and role play helper and helpee roles, with each of you having the opportunity to play each role. Pretend this is the last time you will meet together, and see how effectively you can share your real feelings and concerns. What does this exercise remind you of? How do you usually cope with separation anxiety? Discuss your reactions to this exercise with other people in your group.

Exercise 7.6   This is a Gestalt exercise to put you in touch with feelings of separation anxiety. Take something out of your purse or pocket that is very necessary to you (your wallet, your car or house keys). Put that article under your chair, lean back, and close your eyes. Now imagine that you are not going to retrieve that article and that you need it. Let yourself get in touch with the panicky feelings that emerge, and try to experience these feelings for several minutes. Then discuss the exercise with others.

Exercise 7.7   With a partner, identify five to ten interactions you have had in the last week. Describe how they ended. Who or what ended them? What kinds of feelings did you have? These interactions may have been over the telephone, in person, or even by letter. For the coming week, try to become more aware of how you end telephone and personal contacts.

☐ Step 6: Follow-Up

Follow-up entails checking to see how the helpee is doing with respect to whatever the problem was some time after termination has occurred. This is a step that many helpers do not take. However, some work settings require formal follow-up by mail or telephone. And in other work settings where follow-up is not required, many helpers informally do their own follow-up by dropping in on former helpees to say hello and see how everything is or by dropping them a note or giving them a call.

Here again, the helper must distinguish between genuine follow-up and extension of a helpee's or helper's possible dependency. The purpose of follow-up is to evaluate long-term effects of helping strategies and the effectiveness of the helper. Occasionally, follow-up will result in additional helping; in most cases, however, it simply serves as an evaluation of both helper and helpee. It is a form of recognition that both parties can appreciate, in that it can communicate genuine caring and interest.

I generally telephone former clients six months to a year after termination to let them know I've been thinking of them and am wondering how they are. I allow enough time for them to fill me in on developments in their life, and I check to see if they have been able to use what they learned from the helping relationship in other aspects of their lives.

Now that we've briefly discussed each of the six steps of the strategy stage of the helping relationship, we'll do an exercise that reviews all six of them.

Exercise 7.8   The purpose of this exercise is to help you recognize the different steps of the strategy stage. Which step do you think each of the following numbered excerpts represents? The answers are at the end of the chapter.

1. **Helper:** Hello, Mr. Witkins. This is Mr. _____ calling. How are you?
   Client: Oh! . . . I'm fine, thank you. How are you?
   Helper: I guess you're surprised to hear from me. I've been thinking about you and wondering how everything is going.
   Client: Well, everything is real O.K. I've been working since March and I really like it . . . in the accounting office of _____ Associates. The hours and pay are better than the last job, and I'm home to help out more after supper with the kids.
   Helper: You and your wife getting out together more now?
   Client: Oh, yes. We make a point of going out alone once a week. And we're beginning to have company in now, too. Things really have been better since I found the new job. And I have to thank you for seeing me through that rough period.
   Helper: Well, I'm glad I called. Hope things continue to go well for you. Regards to your wife.
   Client: Thanks. Appreciate your calling.

2. **Client:** I'm really in a bind. There's no place for the old man to go. We can't afford a home, and Jim's brother is out of the country, and I'm just stuck, that's all.
   Helper: Let's focus on what we can do to make this situation tolerable for you and your family. Since, as you say, your father-in-law's residing with you is a given . . .
   Client: Yes, I need to learn not to react so much, not to get so upset every time he says or does anything. I know it's not good for Jim and the children, but I just can't help myself.
   Helper: O.K. We'll focus on the home situation before we talk about your idea of getting a job.

3. **Helper:** How'd you do on your last math test?
   Client: I passed it! And my homework assignments are O.K.
   Helper: Do you feel you're meeting the terms of the contract?
   Client: There was one day I didn't do my work. I just went without watching TV. I didn't like that, though, because the next day my friend Jeff told me I missed a great game.
   Helper: What are you going to do about that?
   Client: I'm not going to miss any more.

4. **Helper:** It sounds to me like somebody once told you when you were a kid that it's not O.K. to cry.
   Client: I never cry. Yes, I remember something happened when I was very little, around four or maybe even three.
   Helper: I'd like to try a role-play with you that may get us in touch with that early decision. O.K.?

Client: Sure.

Helper: Now be that little girl and tell me what happened.

Client: I left my doll carriage on the driveway and my dad ran over it when he came home that night.

Helper: Go on.

Client: I cried . . . and he came up to me and hit me . . . and he said I'll show you what happens to little girls who leave their doll carriage on the driveway and cry. And he took the doll off the driveway—it was a raggedy type of doll—and he tore her apart and threw her away.

Helper: And you decided never to cry again.

Client: *(crying)* Poor little girl.

Helper: *(throwing her a large cushion)* This is your dad. Be that little girl and tell him what you want to tell him.

5. **Helper:** We have only two more sessions together. I'm wondering how you feel about that.

Client: Well, in a way I'm glad, and in a way I'm not sure. I think I'm doing much better. I don't get stoned anymore. But every now and then, I get scared . . . and lonely. I think I'll take your suggestion and join that group you were telling me about.

Helper: We can use the next two sessions to talk more about that, if you'd like.

6. **Helper:** What often works in cases like yours is what we call *systematic desensitization.* Have you ever heard of it?

Client: No.

Helper: It's a form of counterconditioning. I teach you how to relax each of your muscle groups, and when you're able to master that, we begin to imagine all aspects of flying while you're relaxed. It's impossible to be anxious and relaxed at the same time, and this is a technique that can be used whenever you're anxious. I have a paper you can read about it.

Client: How long does it take?

Helper: It usually takes about three sessions for relaxation, and then somewhere between ten and fifteen sessions for desensitization. I often use other strategies along with desensitization, but we can see as we go along.

Client: I'm willing to try.

Exercise 7.9   Select a partner with whom you have already established some degree of trust. One of you will be the helper and the other the helpee. Work through the following steps; then reverse roles and start the steps again.

1. Helpee suggests a problem to work on.
2. Using responsive listening, the helper explores and clarifies the problem with the helpee.
3. The helper restates the problem in terms that are clear, goal directed, and acceptable to the helpee.

4. The helper and helpee proceed to break down this problem into a specific goal statement.
5. Next, the helper aids the helpee in identifying the psychological needs of both parties in this helping relationship (for example, need to be liked, need to be successful, need to control).
6. The helper suggests at least three different helping strategies that will meet the previously identified psychological needs of the helpee.
7. Next, the helper suggests at least three different helping strategies that will meet the previously identified psychological needs of the helper.
8. Then the helper, in collaboration with the helpee, determines if any strategies have been suggested that will meet the needs of both helpee and helper. If possible, the helper selects the one strategy that will best meet the needs of both parties in the helping relationship.
9. Once a strategy has been selected, determine who is going to do what, and when and where he or she is going to do it. Determine what the terms of the relationship and the strategy will be and how you will know if and when the problem is resolved.

The purpose of this exercise is to show you that the needs of both parties in the helping relationship must be taken into account when selecting strategies. For example, a helper who has a need to be liked and to refrain from expressing emotion might not select a Gestalt strategy even if that strategy is likely to be most effective for this particular helpee at this particular time. The more one is in touch with one's own values and needs, the more aware one is of the reasons for choosing particular helping strategies.

☐ Case Studies

Three brief case studies will now be presented to show how counseling progresses through the relationship and strategy stages. These three case studies have been selected to illustrate the steps in building a relationship and applying strategies. You will note that the steps flow freely into one another and often you go back one or two steps and then return.

The names and identifying information have been changed in these case studies to protect the clients' identities.

The Case of Ms. S.

Ms. S., age 48, was referred to this counselor by her counselor at the outpatient clinic of a state hospital because her counselor was leaving the state. Ms. S. had been in and out of treatment for ten years, although she had never been hospitalized. She was diagnosed as being depressed with phobic reactions. For ten years she had refused to leave her home except with her husband, a truck driver for a large baking firm. Thus, all her marketing, shopping, and other errands were done with her husband, who worked an early shift and was home by 4 P.M. Ms. S. lived in a lower-middle-class neighborhood with her mother (age 82), her husband (age 52), and her 22-year-old retarded son. At this particular time, a crisis had arisen because Mr. S. was being transferred to another route, which would involve him in work all day and not leave time for marketing at Ms. S.'s favorite small markets.

Session 1

Mr. S. drove Ms. S. to the session and arrived a few minutes late. He waited outside in the car. Ms. S. apologized for being late, took a while to get comfortable, and then sat primly with her hands folded on her lap. Because she seemed to expect to be questioned and because I already had background information, I asked her to tell me about what she had been doing all day. *(initiation/entry)* Very slowly, she told me about her sewing, her cleaning, and the time her son and mother took. Her great joy and pride in life was her cooking, and she was very fussy about the foods she bought and what she cooked. She refused to shop in supermarkets; she had known the vendors where she marketed for years and could depend on them for fresh, high-quality produce, cheeses, and meats. On Sundays the highlight of the week occurred: her daughter and son came for dinner with their families.

(Counselor's comments: "Ms. S. is neat, nicely groomed, and pleasant to talk with. She does not express much animation, and she talks very slowly and carefully. It took about an hour to get the preceding data.")

Session 2

Ms. S. continued to talk about her family and cooking. She described in great detail the past Sunday: the food and the sayings of her grandchildren, who were quite young. *(clarification of problem)* Toward the middle of this session we began to talk about her fear of leaving the house. She could not remember the onset of this fear and she could not clarify what she was afraid of. She seemed to think it happened gradually and that it was nothing to get upset about. However, when I reminded her that her husband was starting a new route in a few days, she became agitated (her fingers started to play with each other and she licked her lips a lot). She then explained that she lived about eight blocks, less than a mile, from the markets and that, if her husband came straight home from work and there was no traffic, maybe they could still make it. I asked her if she thought she would be physically able to walk that distance, and she replied that sometimes on spring and fall weekends she took walks longer than that with her husband and their son.

Session 3

(Structure/contract) After a few minutes of being caught up on Ms. S.'s household chores, I asked her if she would be interested in learning to be able to walk to the markets alone to get her food. *(exploration of problem)* She said she didn't think she would be able to do that, and we spent the entire session talking about the

pros and cons. *(possible goals)* I suggested that we take no more than ten sessions to try it out and see if we could together enable her to overcome her fear. We went over and over the consequences of her not being able to market where she wanted to, of her husband's new route, of maintaining the status quo. Toward the end of the session we returned to less anxiety-provoking events, and she proceeded to tell me about some outfits she was making for her granddaughter.

Session 4

This session began with some more relationship discussion about my perceptions of and feelings toward her and resulted in her expressing pleasure at our meetings. *(mutual acceptance of goals)* We then talked about her agreeing to focus on the behaviors of being able to go out of the house alone and walk to the market, shop, and return home. *(planning of strategies)* She agreed to give "my way" a chance, and I described a step-by-step behavioral procedure to her whereby I would come to her house each day for the next 12 days; if she was able to meet the directives for the day, I would take her marketing and/or take her to her daughter's house for a visit. When she agreed, we asked Mr. S. to come into the office, explained the procedure to him, and asked him not, under any circumstances, to market for Ms. S. or take her marketing. He agreed and Ms. S. agreed (but with reservations).

Intervention Period

The following 12-day period consisted of graduated steps. A summary of these steps follows.

Day 1: (Use of strategies) Ms. S. put on hat and coat and came to front door and walked downstairs. I said that she did not have to go any farther than that, and I asked her if she wanted to go marketing with me. She said no, she was too nervous; but she was pleased when I offered to go to the bakery for her and get her some of her favorite pastries (which Mr. S. had told me about).

Day 2: Ms. S. was able to walk to curb with me. During this walk, I talked calmly to her, urged her to take long, deep breaths. When we got to curb, we returned to house, and I went to the butcher for her and picked up her order.

Day 3: The goal was to walk down the block to where my car was parked. When we got to the next house, Ms. S. said she was dizzy and wanted to return home. We did so. No reinforcement. I left immediately, not wanting to reinforce her with my company.

Day 4: Ms. S. expressed concern about letting me down yesterday. As I reassured her, we left the house, and we were halfway down the block before she realized where we were. She was pleased and surprised and, as we were by my car, she agreed to be driven to market. She said she was out of produce and she had been "worried sick" about how she would manage.

Days 5–8: Each day, Ms. S. was waiting for my arrival with her hat and coat on, and each day we extended our walk until day 8, when together we reached the market, filled the cart with groceries, and walked home together. This called for a celebration, and we called Ms. S.'s daughter, who came over for a visit, bringing the favorite grandchild. Ms. S. reported that she was so busy talking that she had not been nervous or afraid, even when crossing the streets.

Day 9: (Evaluation of strategies) I waited for Ms. S. halfway to the market. She arrived at the appointed time, and we proceeded to the market together. She had a longer list today, as it was her son-in-law's birthday.

Day 10: Ms. S. was able to come alone to the market, where I met her. We returned home together, and her daughter joined us for lunch. We agreed to add two more days to this period.

Day 11: Following a weekend, I phoned Ms. S. and told her that I would be at her home waiting for her when she returned from the market. She was about ten minutes later than I, and she explained that the produce trucks were slow in being unloaded. She was out of breath, but in good spirits and very busy bustling about the kitchen.

Day 12: Repeat of day 11.

Sessions 5 and 6

(Evaluation) These were short sessions held in Ms. S.'s home (her husband was no longer able to drive her to my office). We discussed what had happened and her feelings. It was very hard for her to talk about her feelings, but she appeared to be O.K. and had much to report about her family and cooking.

Sessions 7 and 8

(Evaluation/termination) These sessions occurred at two-week intervals and consisted of checking things out and discussing forthcoming termination. Ms. S. was continuing to do her own marketing, although she did not leave the house for any other reason, except with her husband.

Session 9

(Termination) This session occurred three weeks after session 8. We talked about what we had accomplished, and I asked Ms. S. whether she thought she might want to go out alone for purposes other than marketing. She wasn't sure about that and seemed reluctant to discuss it. As we had fulfilled our contract with each other, I did not pursue the matter.

Session 10

(Follow-up) This session occurred six weeks after session 9. Ms. S. was still totally involved in her family, household, and culinary efforts. She accepted her husband's changed route now that she could market herself, but she had not made

any attempts or expressed any interest in expanding her independence out of doors.

The objective of being able to leave the house to market was accomplished. At the time of follow-up, there was no generalization of this to other situations, but it could be that there was no reason or motivation for Ms. S. to go other places alone, since her family readily came to visit her, as did neighbors and friends. No attempt was made to resolve depression or underlying causes of phobic behavior.

The Case of Rory

Rory, age 17, was referred to a community counselor by the school drug counselor after he was expelled from his senior year of high school because of drug use. Rory had not been in trouble before high school, although he had been placed in a special program for behavior problems and slow learners after one year in high school. He was such an appealing person that teachers and parents found it easy to forgive his misbehaviors and "give him another chance." His behavior problems before the drug usage were more passive than active—he would not hand in assignments, would be unresponsive to adults, and would engage in foolish activities with other kids that would disrupt the class. Finally, when found dealing drugs to other students, he was put on probation by the courts and ordered to seek therapy. Rory's school record indicated average intelligence and no specific learning disabilities. His personal appearance was attractive and friendly. He had a cute smile and a twinkle in his eye and he was easy to talk with. Rory was the youngest by five years of eight sons in an Irish-American family; his parents were retired and in their late 60s at the time of this referral. Mr. Wilson had been a mail carrier and Mrs. Wilson had worked in a nursing home as an aide. Two of Rory's brothers had attended college, four were married, and all were self-supporting. Several of his older brothers had had drinking problems while in high school and were viewed retrospectively by their parents as "tough kids."

Session 1

Rory and his parents attended the first session together. Rory talked about how boring school was, how much his parents nagged him to be like his brothers and "make something of himself," and how the only fun he had was going out with his friends. Mr. Wilson expressed his frustration about not being able to get Rory to do anything—when he was younger and physically stronger, he was able to "make his boys behave," and he would not tolerate Rory's sassing back and making his mother so unhappy. Rory never helped around the house and didn't talk to his folks at all. He was moody and sullen. Mrs. Wilson expressed dismay at her failure to keep peace in the family (Rory and Dad were always arguing) and said she was very worried about her dying mother. Both parents described their physical ailments at great length—they were worried about their health and worn out by their parenting responsibilities. The counselor noticed that while they were talking, Rory tuned out. He stared at the ceiling with an impassive expression on his face. All three members agreed that Rory came and went as he pleased no matter what restrictions his folks tried to impose, and that they were at an impasse. The counselor commented that while there was a lot of frustration in this family, there was no anger and rage, but rather a great deal of caring and concern. *(Initiation/entry/joining)* Perhaps each person's frustration and depression came from so much caring. At the end of this session, it was agreed that Mr. Wilson would bring Rory to an individual session (part of Rory's probation included the lifting of his driver's license). In the interim, the counselor agreed to look into community-based programs for Rory's drug rehabilitation and schooling, and the parents and Rory were directed to look into the school's own alternative program.

Session 2

The counselor met with Rory alone to develop a relationship and make some immediate drug rehabilitation and educational plans to deal with these entry problems. During this session, Rory talked more openly about how hard it was to be the only kid in the house, how all his friends had younger parents, how his brothers were also on his back because they heard from his folks about how bad he was, and how only his friends understood him. He acknowledged that he smoked pot every day, usually before, during, and after school. He did not seem upset that he had been caught and thrown out of school. He did seem to be interested in getting help but he seemed to be puzzled about taking any kind of responsibility for himself. His behavior appeared apathetic and depressed. The counselor pursued Rory's thinking and past patterns of helping himself when he got into trouble *(clarification of problems)* and helped him to talk more about his feelings about his family and his wanting to be liked and supported. At this stage, Rory did not seem to take responsibility for himself or his problems and seemed content to let the drugs and others make him feel better.

Sessions 3–5

(Deepening relationship, exploration of problems and possible goals) The focus of these sessions was to engage Rory in the helping relationship so that he would become more actively involved in making his own life decisions. During this time, he and his parents visited several possible program sites and a family contract was developed regarding his going out and his use of daytime hours since he was not in school.

Rory agreed to attend Narcotics Anonymous meetings (also a term of his probation) as well as to keep his appointments with various resource people. He reported less arguing at home and by the end of session 5 *(structure/contract)* had worked out some objectives with the counselor: (1) to regularly attend NA meetings; (2) to attend the alternative school program if he was admitted; and (3) to regularly meet with the counselor. He was intrigued by the counselor's suggestion that all of the brothers who lived in the area attend one meeting with him without the parents, and he even offered to make the arrangements *(planning of strategies).* This was the first time he had really actively engaged in the helping process. Other strategies suggested were the use of limited behavioral contracts, Gestalt dialogues, cognitive restructuring, and reality-therapy questioning. For example, Rory was directed to stop before smoking and ask himself two basic questions: (1) Do I know what I am doing? (2) Is this what is best for me?

Session 6
This session was another planning session in that the counselor helped Rory prepare his application materials for the alternative program and rehearsed for the interview by role playing. At the end of the session, Mr. Wilson, who would wait out in the car during the sessions, was asked to come in to share his perceptions of how things were going at home. He reported that the fights were fewer and that while he was worried at how much time Rory was spending lying on his bed listening to his tape player with his earphones, he was able to keep his worries to himself. He still talked about how worried "the missus" was. Rory had told him about the forthcoming session with his brothers and Mr. Wilson thought it was a fine idea and did not seem to be upset that he and his wife were not invited.

Session 7
(Implementation of strategies) This was a critical session in that four of Rory's older brothers came to the session. At first, Rory sat as far away as possible and seemed to be very wary of what was going to happen. As the counselor encouraged each brother to share with Rory what it was like for him growing up in this family and how he worked out his problems, Rory became more alert and interested and joined in the conversation. This one-and-one-half-hour, very emotional session ended in a group hug with the brothers openly expressing their concern and their availability to Rory. Rory left this session with a grin and looking happier than ever. The purpose of this session was twofold: (1) to assess Rory's difficulties within a larger family system context, and (2) to obtain support and modeling for Rory.

Sessions 8–30
(Use of strategies) Rory was admitted to the alternative program and was able to abide by the contract (daily attendance, no drug use, compliance with assignments) with the program administrator. This was a small, structured program individually designed on a local university campus several miles from the high school where Rory had gotten into so much trouble. At first, he would try to return to the high school after hours, but he was directed by his probation officer to stay away from the high school. Gradually, he was weaned away from some of the kids at the high school who had not been a good influence on him. While his drug usage did not disappear, he did curtail it so that it did not interfere with his school attendance. During these sessions, Rory developed a greater attachment to the counselor and would bring in his schoolwork to show her as well as talk about his friends and his family with much more understanding. Person-centered and cognitive-behavioral strategies predominated, although occasionally Gestalt dia-

logues were used to heighten Rory's self-awareness. He became involved with a new girlfriend in the alternative school program and they helped each other to abide by the program contract. He had more contact with his brothers and he was able to become more caring to his parents about their health and family concerns. A great deal of time in these sessions was devoted to planning for his graduation from high school and post–high school life. When his girlfriend became pregnant and decided to have and keep the baby, he responded very responsibly and spent a lot of time asking the counselor for information and resources. He read books about pregnancy, took his girlfriend to the prenatal clinic, and sought an after-school job in a warehouse so as to save some money. Occasional family sessions indicated that while Mr. and Mrs. Wilson were upset about the pregnancy, they were supportive of Rory's responsible behavior and included his girlfriend in their family. His brothers followed through more on their agreement to be available to Rory, and Rory learned to initiate contact with them and not just wait for them to make the first move.

Many of the helping strategies used involved decision making—brainstorming options, collecting information, and planning what to do. When Rory realized in April of his senior year that he would graduate, he decided to enlist in the marines. Because of his drug record, he was unable to do so but he was able to get into the army. Luckily, he had a very supportive and encouraging recruiting officer who helped him over some of the rough application hurdles.

Sessions 31–40

(Evaluation/termination) These sessions took place between Rory's high school graduation and his induction into the army. Rory was saving money, hoping to be able to return from basic training for the birth of his baby, and looking forward to the structure of the army and opportunities for finding a vocation. He was able to talk about his anxieties and how it would feel going away from home for the first time. In preparation for the army, he had stopped smoking pot and gotten himself into better physical shape. His whole self-image had changed and he and the counselor talked about and role played saying good-bye, both in the counseling and with family and friends. Rory felt very good about himself and his progress and said that he realized that he had to think more for himself and not be so easily influenced by others. He understood his parents better and was looking forward to taking care of his girlfriend and baby. The actual termination session was emotional and sad for both Rory and the counselor. It included a review of decision-making and problem-solving steps and of specific ways that Rory could help himself feel better about himself.

Three-Month Follow-Up

This occurred by telephone when Rory returned for the birth of his child. He reported that he was doing very well in the army and that he had remained drug-free and had made good friends, but was primarily focused on his girlfriend and the baby, so most of his free time was spent writing letters home. He had been homesick but he was kept so busy, it wasn't as bad as he had feared. He hoped to save enough money to get married within a year.

Two-Year Follow-Up

Before being sent to the war in the Persian Gulf, Rory returned home and requested a session with the counselor to thank her for "standing by him" but, more important, to ask her if his wife could come see her while he was out of the country. Rory wants to remain in the military. He is finding the structure and opportunities positive experiences and he feels more successful and energetic than he ever has. He no longer feels as if he is "just going along."

This case involved both individual and family counseling sessions. The systems perspective enabled the helper to understand Rory's need for structure and support in his family and at school and to help Rory change the way he related to his family so that they could simultaneously change their ways of relating to him. The steady, continuous support and practical skill building that Rory received during the counseling helped him achieve the primary objectives of (1) completing his high school education and (2) freeing himself from his drug dependency. This was a positive experience for Rory because he was lucky to have a family with the capacity to be caring and supportive and because he lived in a community with available resources. It was also a positive experience for the counselor in that there was so much raw material to work with.

The Case of Martha

Martha, age 30, entered counseling to find out whether or not she wanted to have a baby. She'd been married for eight years and reported a relatively happy marriage. She was an elementary school teacher with many activities and interests, and Bill, her husband, was a middle-level research scientist in an industrial laboratory. They had never tried to have a baby; what precipitated this entrance into counseling was a chance remark by Martha's physician during her annual check-up to the effect that, if Martha and Bill were going to have children, they ought to think about getting started.

Session 1

Martha arrived early for her appointment and was eager to talk about herself. *(initiation/entry)* She told me that she was the second of four daughters from a traditional Jewish middle-class background. She recalled a relatively placid, happy childhood, although she remembered a lot of squabbling among the sisters. Her lingering impression was of an all-female, matriarchal home. Her father seemed almost nonexistent and Martha was vague in her references to him. Martha was an above-average student, although not as academically gifted as her older sister. She was easygoing, anxious to please, and easily led by others.

Her older sister went to a prestigious college, but Martha attended the state university, where she maintained a B– average and dated moderately. She met Bill during her senior year, and they were married right after graduation. She had been teaching fourth grade ever since, and said she liked teaching and maintained a well-organized, well-run classroom. She did express disdain at parents and teachers who didn't know how to "control and discipline" children. Martha had supported Bill while he studied for his master's degree. They now lived in a two-bedroom apartment in a suburb. When they got married, they had not discussed having children, because they had decided that was a long way off.

Martha described Bill as quiet, introverted, a loner. They didn't see too much of each other because she had her women's group, her art class, and her card group during the week. However, on Saturday nights they either had company or went out. Sundays were for visiting families, who lived nearby.

Martha had been "shocked" by her doctor's comment. She'd had no idea she'd have to start thinking about children. A friend suggested she come for counseling, and she eagerly accepted that suggestion.

(Counselor's comments: "Martha is attractive, bright, and very outspoken. She was eager to talk and speaks rapidly and egocentrically—practically every statement begins with 'I.' Near the end of the session, I asked her about Bill and she seemed genuinely startled. She does not know how Bill feels about having children, but she has decided that she needs to figure out how she feels first before she discusses it with him. She definitely does not want him to come to counseling with her.")

Session 2

Martha arrived early again, but she seemed a little subdued, her rate of speech was slower, and she was more reflective. *(clarification of problem)* She described her eight years of marriage as if she were relating an article from a women's magazine, with little affect. She talked about their different interests—she liked to keep busy, always doing things and doing them as perfectly as possible, liked to be with people, go to cultural events. Bill's idea of fun was reading a mystery novel or watching TV. While he was watching TV during the evenings she was at home, she was usually reading and correcting papers. She had pretty high expectations of her students and wrote long comments on their papers.

We discussed her feelings about children: no one she knew was happy with his or her children; too much work; she'd be trapped and tied down; she'd hate to give up teaching, yet she wouldn't dream of working while her children were young; it seemed like such a burden. The only positive reason she could think of for having children was that "it's the thing to do" and that "it would please our parents." She said that Bill didn't really care—"he wants what will make me happy."

The session concluded with Martha agreeing to keep a diary containing all the thoughts, feelings, and observations she had or made about children during the week.

(Counselor's comments: "The fact that she's come for counseling is the only behavioral evidence that there is a positive side to her feelings about having children. There seems to be some anger and fear underneath the bright exterior.")

Session 3

Martha produced her diary for the week, which elaborated her ambivalences. *(clarification)* On the one hand, she was afraid that if she decided not to have children she would regret it when it was too late, and on the other hand, she had many "shoulds" in her head that she was afraid of, such as "I should have children because it is expected of me"; "If I have children, they should be perfect"; "If I have children, I should stop teaching and stay home and be the perfect mother." We spent much time talking about these beliefs and what they meant to Martha— the feelings behind them. She had great difficulty experiencing her own feelings. She could talk about, but not experience them.

(Structure/contract) We decided to set up seven more sessions, one per week, to see where we could go. Martha agreed to continue her diary and to read some books that I'd selected for her.

(Counselor's comments: "Slowly but surely, she's becoming a little less guarded and more relaxed. She's not trying to impress me as much. Less talk about all her activities and achievements, and today she expressed some genuine feeling [compassion] for one of her students.")

Session 4

(Exploration) We talked about where Martha got some of her beliefs and what kinds of things occurred in her childhood to reinforce them. As she talked about her mother expecting her to live up to her older sister and to keep on trying harder, she began to get in touch with some anger. *(use of strategies)* I explained the TA ego states to her, and she decided she had definitely been an "adapted child." Using some Gestalt strategies, I encouraged her to experience the anger, and she did remarkably well for the first attempt.

Sessions 5 and 6

(Exploration and possible goals) We decided to focus on Martha's feelings about herself and her self-understanding, rather than on the decision making about children. *(mutual acceptance of goals)* It became apparent that there was much material to work through before a decision could be made. Martha decided that she really didn't like herself very much and that she hid behind lots of activities and fast talk.

(Counselor's comments: "She seems somewhat relieved, although anxious, that she doesn't have to play 'perfect' games anymore and that she can say and do what she feels like here.")

Sessions 7–10

(Planning of strategies) In these sessions, I explained how we could use some rational-emotive, Gestalt, and transactional-analysis techniques to help Martha get in touch with and understand her feelings and beliefs. She agreed and easily got into dialoguing, developing fantasies, and so forth.

Martha came into the ninth session disturbed by the anger she was feeling toward her family. She found now she was feeling angry every time she spoke to or thought of them. *(use of strategies)* I emphasized that it is O.K. to be angry and that she would be able to work it through once she allowed herself to feel it. During a dialogue session with her imagined mother, some anger spilled over to her father and Bill, both of whom she sees as "passive," "almost not there." We spent the rest of the ninth and all of the tenth sessions dealing with the anger toward men. *(recontracting)* At the end of the tenth session, we agreed to meet for five more sessions.

Sessions 11–14

We continued intense exploration and working-through of anger and the early script decision wherein Martha had vowed always to try harder.

(Use of strategies) Each session ended with a behavioral contract for the interval between sessions, such as talking to Bill about a particular feeling or doing something fun and enjoyable. The contract after session 13 was to tell both families that they would not visit that Sunday because they were going to spend the day together alone. Session 14, Martha came in glowing: she had had a wonderful day with Bill, and they had really talked for the first time in years. *(evaluation)* She said she was seeing and hearing things she had never seen or heard before, one such thing being Bill's caring for her and wanting her to stop running around so much and to pay more attention to him.

Session 15

This was the final session contracted for. *(termination)* Martha said that she wanted to feel her own way from here on and that she believed that over the next few months she and Bill would be able to decide about having children. She felt good knowing that whatever they decided would be fine for them, and she agreed that she was now able to balance the scales a bit by seeing and looking for people who enjoyed having children.

Three-Month Follow-Up

Martha reported over the telephone that although they still had not made a final decision with regard to children, they were getting there. She reported that she was able to continue to refute her "crazy" beliefs and that she was able to relax and to enjoy Bill's company more. She thought that one day they would both want to come in for some counseling together, as she knew there were marital issues they could be helped with.

You will note that I use bibliotherapy (the assigning of pertinent reading material) and weekly homework assignments regularly. I find that these behavioral strategies go hand in hand with the more experiential strategies used in sessions. In this case, it was important not to rush into decision making but to take the time to explore the underlying dynamics of the presenting concerns.

☐ Summary

In this chapter, we have discussed the six steps of the strategy stage: mutual acceptance of defined goals and objectives; planning of strategies; use of strategies; evaluation of strategies; termination; and follow-up. We stressed the importance of a theoretical framework within which helpee problems are defined. The goals and objectives of the helping process are derived from this problem definition, and the helper's theoretical framework helps in the selection and application of strategies.

Examples and exercises were presented to help you understand the steps. We particularly emphasized the termination and follow-up steps, as they are too often neglected in favor of the relationship development steps. We presented three case studies to illustrate the steps of the human relations counseling model. Remember that these steps are not necessarily distinct and discrete. Each case is different, and flexibility and modification are encouraged. Further, the effectiveness of the strategy stage depends on the quality of the relationship stage, as demonstrated in the case studies. Thus, the steps in applying strategies can be viewed only within the context of an empathic helping relationship.

☐ Exercise Answers

Exercise 7.4   1. c and d; 2. a, c, and also d; 3. c, d; 4. b, d; 5. a, b
Exercise 7.8   1. follow-up; 2. mutual acceptance of objectives; 3. evaluation; 4. use of strategies; 5. termination; 6. planning of strategies

☐ References and Further Reading

Corey, G. (1991). *Case approach to counseling and psychotherapy.* (3rd ed.). Pacific Grove, CA: Brooks/Cole.

Corsini, R. J. (Ed.). (1991). *Five therapists and one client.* Itasca, IL: Peacock.

Okun, B. F. (1990). *Seeking connections in psychotherapy.* San Francisco: Jossey-Bass.

Ward, D. W. (1984). Termination of individual counseling: Concepts and strategies. *Journal of Counseling and Development, 63,* 21–26.

Wedding, D., & Corsini, R. J. (Eds.) (1989). *Case studies in psychotherapy.* Itasca, IL: Peacock.

EIGHT

Crisis Theory and Intervention

Today helpers are working more and more frequently with people who are experiencing some type of crisis; in other words, helpers are *intervening* in crises. Crisis intervention involves the short-term use of specific skills and strategies to help people in crisis cope with the turmoil resulting from specific emergency situations or events. Crisis intervention is active, direct, and brief, and occurs shortly or immediately after the crisis becomes evident. It ranges from immediate contact and support to arranging for intensive treatment or therapy. Crisis intervention is an approach to helping relationships that is usually distinct from the counseling model previously developed in this text. However, it requires the same communication skills for establishing a relationship and clarifying and understanding the crisis, and it can use many of the problem-solving strategies of the human relations counseling model.

More recently, crisis theory has been broadened to include trauma or disaster theory. Our present world is linked by the news media and rapid communications, which spread the effects of natural and technological disasters such as airplane crashes, terrorist attacks, violence, earthquakes, hurricanes, tornados, and war. Crisis intervention may focus on one or several victims and may involve conventional one-to-one helping, whereas trauma or disaster intervention requires coordinated team efforts to deliver services to larger groups of sufferers and the use of multiple modalities of helping.

The purpose of this chapter is to familiarize you with basic crisis theory and with the application of helping strategies in basic crisis intervention.

☐ What Is a Crisis?

A crisis is a state that exists when a person is thrown completely off balance emotionally by an unexpected and potentially harmful event or a difficult developmental transition. The major difference between stress and crisis is that a crisis is limited, whereas stress can be ongoing. Crises are not usually predictable or expected, and it is this unexpectedness that can intensify the reaction to crises. When we are in crisis, we feel a loss of control and power over ourselves and the course of our lives. Common terms used to describe the results of a crisis are *disequilibrium, disorientation,* and *disruption.* It is the intense emotional experience of these states that creates the crisis. Common feeling responses to crisis include apathy, depression, guilt, and loss of self-esteem. People in crisis find that the ways that they solved problems and coped with difficulties in the past no longer work, and they become more and more upset and frightened.

When we talk about crises, we are referring to people's emotional reactions to a situation, *not* the situation itself. Therefore, crisis intervention helpers work with a person's perceptions and judgments of the crisis, not with the event itself. If a person comes to you in crisis because of an accident, you deal with that person's feelings and thoughts about the accident, not with the accident as an isolated event. How one responds and reacts to crisis depends on one's past learning and experiences (how one has reacted to meeting minor crises throughout childhood and adolescence), one's lifestyle, and one's life philosophy.

For example, a volunteer rape counselor reported that two married women in their 40s with similar educational and socioeconomic backgrounds came to her for crisis counseling on the same night. The situations were so alike that the police suspected that both women were assaulted by the same person.

One victim collapsed hysterically and required intensive care and counseling, missing work for two weeks. The other victim required factual information about police and court procedures before driving herself home (20 miles), having a drink with her husband while she related the details of her assault, going to bed, and reporting for work the next morning. Neither reaction is "better" or "more normal" than the other. These two women had different histories, philosophies, and coping and defense mechanisms. They needed different amounts of time and types of help in working through their feelings about their crisis.

Exercise 8.1   Close your eyes and reflect on the last loss you experienced. It may have been the loss of a job, a loved one, a pet, a wallet, or whatever. Focus on the feelings you experienced as a result of that loss, not on the loss itself. Remember where you were, with whom, and what everything looked, sounded, and felt like. How did your distress show itself? See if you can remember all of your feelings, their physical expressions (for example, tight stomach, clenched fists, palpitations), and the thoughts you had about the event and your feelings. Were those feelings familiar to you? How long did you feel the acute pain from loss? What coping strategies do you remember using to pass through this state? After you have remembered these details, pick a partner and share your findings. See what you can learn about your own way of dealing with loss and how it may differ from someone else's way.

☐ Kinds of Crises

There are six generally accepted classes of emotional crises:

1. *Dispositional crises*: This type of crisis can ensue from a lack of information, such as not knowing which job to take, what type of medical referral to seek for a particular symptom, what one's options are about living arrangements, whom to ask for what.

2. *Anticipated life transitions*: These are normative, developmental crises that are fairly common in our society. They may result from midlife career changes, getting married, becoming a parent, divorce, the onset of chronic or terminal illness, or changing schools.

3. *Traumatic stress*: These crises result from externally imposed stress situations that are unexpected, uncontrolled, and emotionally overwhelming. Examples are rape, assault, sudden death of a loved one, sudden loss of job or status, sudden onset of illness, accident, war.

4. *Maturational/developmental crises*: Most of us experience these general crises as we pass through our life stages. They may reflect issues of dependency, value conflicts, and sexual identity, or our capacity for emotional intimacy, our response to authority, or our level of self-discipline. Usually these crises surface in relationship patterns or at crucial transition points in our development. Examples are the repeated loss of jobs because of an inability to get along with supervisors, the intense homesickness or depression of college students away from home for the first time, and midlife crises.

5. *Psychopathological crises*: These are emotional crises precipitated by preexisting psychopathology. In other words, one's psychopathology significantly impairs or complicates the way one deals with a situation, inflating it to crisis proportions.

6. *Psychiatric emergencies*: These are crisis situations in which one's general functioning is severely impaired and one is rendered incompetent or unable to maintain responsibility for oneself; in other words, one is dangerous to oneself and/or to others.

As we look at this classification scheme, we see that crises fall into one of two major categories: they are either developmental, in that they have to do with growth and passing through various life stages; or they are situational, in that they are the result of internal and/or external stresses. In addition to helping us understand the nature of crises, the classification scheme helps us to put crises in perspective so that we can determine the best means of immediate crisis intervention.

Exercise 8.2   Using the preceding classification scheme, place the following crisis situations where you think they belong. When you have finished, compare your answers with those of your classmates and discuss the reasons for your answers.

Bad drug trip
Alcoholic binge
Suicide attempt
Acute bereavement over loss of a parent
Loss of job
Discovery of unwanted pregnancy
Discovery that spouse is involved in extramarital affair
Severe marital fight
Abandonment by spouse
Car stolen
Transferring to new high school midyear
Emergency appendectomy in foreign country
Seeing pet run over by hit-and-run driver
Being rejected by college of choice
Being beaten by spouse
Discovering that child is mentally retarded
Finding home vandalized and burglarized

A disaster is a traumatic stress that involves a group of people and organizations. In fact, the consequences of a disaster may reach across the world and the effects do not have defined limits. One important feature of a disaster is that it is a shared experience with shared meaning and influence on group activities, roles, and relationships. Thus, there is immediate societal support available to victims of disasters, which validates their perceptions and experiences. This is in contrast to the confusion, isolation, and alienation that many victims of individual crisis (such as any type of abuse) experience as they struggle with whether or not the abuse actually occurred and if they have a right to their feelings and claims of victimization.

We now recognize post traumatic stress syndrome (intense guilt, anxiety, and/or depression resulting from repeated intrusive thoughts related to the trauma) as an immediate or later response to both individual crisis and disaster trauma. We are attempting to learn more about this syndrome, to discover why some people recover from the acute event and others experience either the delay of stress symptoms or cyclical or continued chronic symptoms.

☐ Who Deals with Crises?

We can see that different levels of helping might be required in different types of crises. Certainly professional medical help is necessary in psychiatric emergencies, and professional psychological help is necessary in crises involving psychopathology. However, any one of us can help with dispositional crises, and many of us can help with anticipated life transitions, traumatic stress, and maturational/developmental crises.

Police, friends, pastors, family, or physicians are usually the first to be alerted to a crisis. They, in turn, usually call in the counselor or human services worker to help in direct problem solving and working through the feelings associated with the crisis.

Exercise 8.3   Choose a partner for a roleplay between helper and helpee. The helpee should pick one of the crisis situations from Exercise 8.2 and act it out. The helper is to draw out the helpee's feelings and thoughts about this crisis. See if together you can decide how to resolve it. After you have finished, process your thoughts and feelings. What stereotypes and value conflicts did you encounter? How comfortable were you talking about this crisis situation? How well do you think you could deal with others' intense emotion in similar situations? What would you want to do differently the next time around?

☐ Crisis Theory

Crisis theory is based on the pioneering work of Eric Lindemann, who studied the reactions of bereaved families of victims who died in the Coconut Grove nightclub fire in Boston. Lindemann (1944) discovered that crisis usually involves some type of loss that necessitates a grieving (bereavement) period. Part of this grieving includes the expression of emotions and intense distress. This distress may take variable forms, such as tightness in the throat, choking, shortened breath, sighing, exhaustion, lack of strength, digestive problems, insomnia, altered sensitivity, preoccupation with guilt, and disturbed interpersonal relationships. These expressions of grief are acute, have an identifiable onset, and last for a relatively brief time (about six weeks).

Together with Gerald Caplan, Lindemann began a community mental health program based on his discoveries. Lindemann and Caplan believed that people in crisis choose adaptive or maladaptive ways of coping with problems and that the nature of their problem solving will affect their later adjustment and ability to cope. They believed that people can be helped to identify, understand, and master the psychological tasks involved in grieving and posed by crises.

In developing this crisis theory, Caplan described four phases of a crisis reaction:

1. *Phase 1:* The initial phase, in which one experiences the beginning of tension and attempts to use habitual kinds of problem solving to restore one's emotional equilibrium.
2. *Phase 2:* This phase is characterized by an increase in tension, leading to upset and ineffectual functioning when one's habitual problem-solving strategies fail; at this phase, one attempts trial-and-error strategies to resolve the problem.

3. *Phase 3:* This phase is characterized by increased tension, requiring additional helping resources such as emergency and novel problem-solving strategies; if one is successful at this phase, one is able to redefine the problem and resign oneself to it or resolve it.
4. *Phase 4:* This phase occurs when the problem has not been resolved in the previous phases and may result in major personality disorganization and an emotional breakdown.

From Lindemann and Caplan's crisis intervention work, we learn that a person in crisis can be receptive to major change in a brief period of time and can be influenced and helped by others during that period. Relationships with significant others (helper, family, friends) are an important part of crisis intervention. A person in crisis needs as much support and help as possible from whoever is available to give it. Lindemann and Caplan found that adaptive crisis resolution *can* result in enduring positive change.

Current crisis theory suggests that unresolved bereavement from earlier losses (of a person, a relationship, security, capacity, a dream) that may or may not be associated with crisis or trauma (such as physical, sexual, or emotional abuse) affects not only one's later day-to-day functioning, but also one's reactions to subsequent crises. Thus, it is important for helpers to learn about the victim's past experiences with abuse and loss so that helping strategies can be planned that enhance one's style of coping.

Although crisis theory is indeed distinct from helping theory, it is consistent with and influenced by the theories of helping discussed in this text. For example, the extensive research on suicide, which focuses on the relation of current crisis to earlier experiences and of current crisis to birth trauma, birth order, and familial and other interpersonal relationships, shows the influence of psychodynamic theory on crisis theory. The influence of phenomenological theory is evident in the existential approach of crisis intervention, which emphasizes the here and now and the positive, growth-potential force of crisis. If one learns to cope effectively with a crisis, it will facilitate one's problem solving during future difficulties. The use in crisis intervention of supportive relationships with significant others and with an empathic helper, to allow the helpee to express intense feelings without interfering with his or her functioning, also demonstrates the influence of phenomenological helping theory.

The cognitive helping theories are represented in crisis intervention theory in that helpers focus on improving the helpee's cognitive appraisal of the crisis and correcting his or her faulty or irrational thoughts. The cognitive-behavioral approaches contribute an understanding of reinforcement and problem-solving theory. The systems theories enable us to understand the crisis event in the context of one's significant relationships, focusing on both the contributing sources and available helpful resources.

The major difference between helping strategies and crisis intervention strategies is the latter's focus on immediate, time-limited reactions to a specific source of stress. In the next section, we'll examine in more detail how helping strategies influence crisis intervention.

☐ Crisis Intervention

The major goal of short-term crisis intervention is to provide as much support and assistance as possible to individuals and their families in order to enable helpees to regain their psychological equilibrium as quickly as possible. From the crisis theory we have discussed, we can derive six major components of crisis intervention.

1. The focus of crisis intervention is on specific and time-limited treatment goals. Attention is directed toward reduction of tension and adaptive problem solving. The time limits can enhance and maintain client motivation to achieve the specified goals.
2. Crisis intervention involves clarification and accurate assessment of the source of stress and the meaning of the stress to the helpee, and it entails active, directive cognitive restructuring.
3. Crisis intervention helps clients develop adaptive problem-solving mechanisms so that they can return to the level at which they were functioning before the crisis.
4. Crisis intervention is reality-oriented, clarifying cognitive perceptions, confronting denial and distortions, and providing emotional support rather than false reassurance.
5. Whenever possible, crisis intervention uses existing helpee relationship networks to provide support and help determine and implement effective coping strategies.
6. Crisis intervention may serve as a prelude to further treatment.

Lazarus's BASIC ID model may be useful as a comprehensive assessment scheme for crisis intervention helpers. One can quickly elicit information about the impact of the crisis event in the seven modalities to determine which modality requires immediate attention and to determine which strategies best fit the particular client in the particular circumstances.

Exercise 8.4   Social networks can be useful in a variety of helping situations. This exercise is designed to help you articulate your own support systems. Draw an ecogram, or social network inventory, to see who actually comprises your current support network. List in columns on a sheet of paper your immediate family, extended family, neighbors, intimate friends, lover(s), casual friends, work associates, school associates, recreational friends, formal helpers (such as doctors, teachers, clergy), and other community members. After you have completed your list, place the appropriate number(s) from the following list next to each name.

1. Close contact, at least once per week
2. No contact within past year
3. Casual, infrequent contact—only a few times per year
4. Have actually experienced support from this person in the past
5. Know I can *always* depend on this person for help
6. Not sure whether I can depend on this person
7. Know I *cannot* depend on this person
8. Would feel uncomfortable receiving support from this person

Stages and Steps of Crisis Intervention

Although there are many different models for crisis intervention, our human relations counseling model can be modified and used to delineate commonly accepted crisis intervention stages and steps.

Stage 1: Relationship
  Step 1: Initiation/entry
   a. Explore various relationship systems (family, work, peer, neighborhood)
   b. Create ongoing opportunity for helpee to express and ventilate intense feelings
  Step 2: Clarification of problem being presented/assessment of crisis problem
   a. Assess major environmental variables
   b. Determine helpee's perceptions of personal strengths/weaknesses
   c. Determine precipitating events (particularly those of the past 24 hours) resulting in crisis—that is, significant change or loss
   d. Determine reason helpee is seeking help at this time
   e. Determine kinds of problem solving and coping strategies helpee has attempted in dealing with crisis
   f. Assess phase and classification of crisis: Is helpee dangerous to self, to others?
  Step 3: Structure/contract for helping relationship
   a. Inform helpee of what you can and cannot do
  Step 4: Intensive exploration of crisis situation and reactions
  Step 5: Discuss possible goals and objectives as well as time limits of crisis intervention
   a. Reiterate problem focus
   b. Reaffirm time limits
   c. Determine how other people and resources can be used
   d. Clarify who is responsible for what (for example, for drugs, referral)

Stage 2: Strategies

Step 1: Mutual acceptance of defined goals and objectives of crisis intervention

Step 2: Planning of strategies

Step 3: Use of strategies
   a. Cognitive restructuring
   b. Referral
   c. Supportiveness
   d. Assertiveness training
   e. Behavioral contract
   f. Ventilation of feelings
   g. Understanding nature of crisis
   h. Decision making
   i. Systematic desensitization
   j. Gestalt experiments

Step 4: Evaluation of strategies

Step 5: Termination when crisis resolved
   a. Formulate realistic plan for immediate future
   b. Verify that helpee is detached from intense emotional reaction
   c. Confirm that helpee has accurate cognitive appraisal of crisis event and appropriate management of affect
   d. Ensure that helpee is willing to seek and accept help from others when appropriate
   e. Confirm that helpee understands how present crisis can help him or her cope with future events

Step 6: Follow-up
   a. Depending on context of crisis intervention, determine if crisis resolution endured

The strategies listed under stage 2, step 3, are taken from Chapter 6 and are merely suggestions of strategies that can be used for crisis intervention. The point is that any strategy that works for a particular helpee quickly and effectively is valid for crisis intervention.

Brief Therapy

Brief therapy is a problem-focused form of therapy limited to ten sessions or fewer. While there are many approaches, techniques, and philosophies of brief therapy, the four-step helping model developed by Watzlawick, Weakland, and Fisch (1974) is useful in some types of crisis intervention.

1. Describe the problem (or crisis) in concrete behavioral terms: frequency, duration, consequences, situational variables; try to understand the function of the problem, what the payoffs and purposes are.

2. Investigate previous attempts at problem resolution. "What have you done about this in the past?" "How?" "What happened when you tried something?"

3. Obtain a clear definition of the change to be achieved. What needs to happen for the helpee to feel better about this? How much change is the helpee willing to accept in order to be satisfied?

4. Formulate and implement a plan to produce the change. What will happen if the change actually takes place? How will the helpee deal with the consequences of change? What may happen to prevent change from occurring?

The focus of brief therapy is on solutions, not problems, on what works for helpees rather than on what has not worked.

Consider the following case:

Mrs. M., age 35, called the counselor sobbing about her husband's latest episode of coming home drunk and beating their only child, 11-year-old Tina. The counselor arranged to see the distraught mother that same day. In describing her problem, Mrs. M. reported that Mr. M. came home drunk about once every three or four weeks, usually after receiving his paycheck. When this happened, he immediately picked on Tina, using whatever excuse he could find, like her coat being out of place or her books being on the kitchen table. He yelled more than he hit, but this time he had really gone out of control and persisted in his hitting. After most altercations, Tina went to her bedroom crying and Mr. M. fell asleep on the couch. Mrs. M.'s usual practice was to go into Tina's bedroom to comfort her. Until today she had been successful in convincing Tina to placate her father; by the next morning, things usually would have calmed down and everyone would act as if nothing untoward had occurred. Mrs. M. was afraid of her husband and had never mentioned his drinking or treatment of Tina to him. What distinguished this incident was the severity of physical abuse and Tina's refusal the next day to talk to her father, who was acting hurt and bewildered. It took two sessions before Mrs. M. could verbalize the change that she wanted: for her husband to cease picking on Tina when he came home drunk and for Tina to resume friendly relations with her father. After much discussion, Mrs. M. was able to see that if Tina were no longer the target of her father's drunken bouts, Mrs. M. might be, and that she would have to learn assertive behavior to protect Tina and confront her husband. Cognitive restructuring and assertiveness-training techniques were used, and by the sixth session Mrs. M. felt that the situation had changed: she reported meeting her husband at the door the next time he came home drunk and dealing with him directly—and she felt more in control of the situation. Although Mr. M. refused to come in for counseling, Mrs. M. was amenable to attending some Al-Anon meetings and learning more about alcoholism.

Forms of Crisis Intervention

There are two main forms of crisis intervention: (1) hot lines, drop-in centers, and crisis clinics where victims can come in person or telephone 24 hours a day; and (2) outreach counseling, where helpers go to the victim to provide immediate support and comfort as soon as they are notified of the crisis. Crisis intervention settings are usually staffed by volunteers who undergo intensive, short-term, on-the-job training. They may or may not be supervised by professional helpers. In addition to developing crisis intervention centers, many schools and communities are working on crisis prevention.

Hot Lines, Drop-In Centers, and Crisis Clinics

Helpers working through hot lines, drop-in centers, and crisis clinics deal with suicide, drug, runaway, rape, alcoholic, and abortion crises, to name just a few, and helpers are taught specialized knowledge about those issues. For example, a drug counselor knows the names of commonly used drugs, the effects of those drugs and their duration, and how to help a user on a bad trip. The suicide counselor knows how to recognize suicide threats, has studied the facts and statistics (not the myths) about suicide, and is aware of different kinds of intervention, such as providing a network of supporters for the victim during the crisis, helping the victim and his or her family to change behaviors to alleviate the crisis, and helping the victim gain a different perspective on the crisis and the situation precipitating a suicide attempt.

In these settings, helpers sometimes get only one or two chances to work with the victim. Therefore, they must be skillful at establishing empathic relations and providing accurate information and alternative options to the victims, all within a short period of time. Contact is important—whether over the telephone or in person. This contact may involve helpers' working overtime, continuing contact by telephone, or effecting immediate referrals.

Telephone calls must be focused on immediate concerns. Quick rapport is essential to deal with present issues. Hot lines are often more accessible than face-to-face counseling and also afford anonymity to the caller. But there is a lack of continuity due to the transience of the hot line encounter, and there is no feedback to the helper and no follow-up. Also, some helpers find the lack of visual cues disquieting.

The following is an example of a late-night hot line call to a university crisis center.

Caller: Hello? Are you there?
Helper: This is the _____ . May I help you?
Caller: I just need to talk to someone . . . I'm worried about my friend. I think he may be thinking about taking pills.
Helper: You're worried what can happen if someone overdoses on—what kind of pills?

Caller: Well, I dunno. Maybe Tylenol . . . he said that's what he'd take.

Helper: It's depressing and scary when someone is all alone, worrying. Taking too many pills may seem a way to get out of it.

Caller: Yeah. I wouldn't do that, of course, but he might.

Helper: Sounds like your friend needs to talk to someone about how bad he's feeling.

Caller: Well, he called some friends but they weren't home and his folks don't ever give a damn *(lots of anger in voice).*

Helper: So he's feeling lonely and rejected.

Caller: Yes, I guess so . . . actually, it's not my friend, it's me—and sometimes I feel like there's no one who cares enough to listen to me.

Helper: Um.

In this case, the helper spent about 35 minutes on the phone and then referred the helpee to the university counseling center. Follow-up indicated that the helpee did report to the counseling center the next morning.

Telephone skills necessary for this type of work include the ability to use responsive listening to establish quick rapport, and the patience to nondirectively follow the helpee's messages to ascertain the nature and severity of the problem. In the preceding example, if the helper had pushed the helpee to acknowledge that the problem was his rather than his friend's, the helpee might have hung up before help could be obtained. Patience, calm, and the courage to hang in there are necessary attributes for hot line workers.

Outreach Counseling

Sometimes helpers must go to the crisis victim rather than wait for the victim to come to them. This is a relatively new concept in human services and is based on the "visiting nurse" concept, where primary care is taken to the client in the client's own setting. The advantage is that the helper is able to see the victim in context and to draw upon immediately available resources such as family and neighbors. In addition, the helper is able to provide direct assistance in immediate problem solving (for example, finding a sitter for the children in the case of a parental accident, talking down a drug victim on a bad trip in a familiar setting, or arranging for immediate medical care).

Outreach counseling usually involves more time and cost than other types of counseling; perhaps that is why so little exists. The time the outreach worker spends traveling to and from clients is time not used for helping.

Outreach crisis intervention is a crucial response to disasters, such as airplane crashes, earthquakes, terrorist attacks, war. In a disaster situation, outreach counseling involves coordinated teamwork. Immediate attention is provided to the victims, the bereaved (the family or friends of victims), service and support providers (other people in contact with the victims such as the police or firefighters, mechanics, and so forth), and those who just happen to be in the immediate environment. Because of the immediacy of a disaster to media viewers, help is often even necessary for those peripheral to the actual event. For example, during the Persian Gulf war, many television viewers suffered symptoms of distress as severe as those experienced by the ones actually involved in the fighting. While most cities have medical disaster plans, only recently have mental health disaster plans been developed. Organized mental health disaster plans will speed the delivery of services and allow for better coordination and quality of help, thereby reducing the chaos that naturally follows disaster.

The outreach counselor feels comfortable in different settings and is not limited by the clock or site. Some are on 24-hour call and think nothing of accompanying clients to the welfare or employment office or helping them obtain legal, educational, and health care. They engage in recreational and leisure-time activities with clients, know how to break up fights and deal with implosive violence, and have learned to establish trusting relationships in the most suspicious climates. Research, though limited, has shown that the best outreach counselors are those recruited from and trained within the communities where they will work. They are in touch with the mores and lifestyles of their neighbors and can more easily overcome distrust.

Outreach counseling requires unlimited patience and dedication. Crisis intervention is only one part of outreach counseling, which occurs throughout the community—on the streets, in bars, houses, schools, social centers, and on playgrounds. This is an active type of counseling that requires full participation in the life of the community. In addition, empathic listening skills are crucial for establishing and maintaining relationships within the community.

Crisis Prevention

Many educational programs have arisen on college campuses and in communities to help people avoid certain kinds of crises and cope with some of the developmental crises that are inevitable. For example, it is routine today for college freshman-orientation programs to include rape-prevention classes, sex-education classes, drug-education classes, and so on. In fact, some of these classes are now part of the regular curriculum in secondary and elementary schools. Crisis intervention workers often contribute directly or indirectly to these educational programs; needless to say, their experience provides them with invaluable knowledge and information. Community agencies are often instrumental in producing and disseminating literature about potential crises, such as pamphlets on how to avoid assault, on neighborhood watch programs, and on first-aid procedures.

Exercise 8.5   In pairs, role play a helpee and helper during a crisis intake interview. After approximately 30 minutes, each helper should write down (1) treatment goals and (2) favored crisis intervention strategies. Switch roles. Then everyone in the group should compare and discuss his or her evaluations. Possible crisis situations to role play are a bad drug trip, date rape, discovery of an unwanted pregnancy, or loss of necessary financial aid in the last year of school.

Exercise 8.6   In pairs, sit back to back and role play a hot line situation. The helpee can select any type of crisis situation. Remember, as helper, you have just this one time to talk to the helpee. Continue your dialogue as long as is natural. When you have finished, discuss your experience together and then share your reactions with the rest of the class. How did it feel not to have visual stimuli? What kinds of pressure did you experience?

Skills for Crisis Intervention

In crises, helpers' abilities to remain calm in an emergency, to use common sense, and to project self-confidence are important. Helpers rely greatly on their responsive listening communication skills, both to get at the nature of the crisis and its stressful ramifications and to communicate comfort, support, and respect to the helpee.

In addition to responsive listening, physical gestures such as holding hands, putting an arm around the helpee's shoulders, and holding the helpee close communicate caring and concern. Except in the case of loss, it is important to differentiate between empathy and sympathy, as the latter can impede recovery from a crisis by fostering prolonged dependence. One way to avoid fostering dependence is to focus on what can be done about the crisis rather than repeating over and over again the details of the actual crisis situation. In other words, after the initial ventilation period, during which the crisis victim describes in great detail the actual crisis, focusing on future action rather than on what happened is more conducive to recovery. Focusing on the helpee's past and current strengths and positive experiences emphasizes the positive rather than the negative and builds the helpee's faith in his or her capacity to recover.

In the beginning steps of crisis intervention, the helper may judiciously pose some questions to the helpee in order to determine which crisis intervention strategy to follow. In addition to asking questions about the actual event, the helper may ask the following kinds of questions:

What changes have occurred in your life recently, particularly in the past few days?

Have you had any particular difficulties with people who are important to you, like a family member, boss, valued friend?

What kinds of things have you already tried to do about this?

Have you ever experienced these kinds of feelings before? If so, when and what did you do about them?

What do you think you need to have happen in order to get through this?

Who in your life do you think might be most helpful to you at this time?

You can see that the purpose of these questions, in addition to eliciting information quickly, is to engage the helpee in the process of understanding crisis reactions and participating in adaptive strategies.

At the same time that helpers are communicating comfort, support, and respect to crisis victims, they may need to take some kind of direct action. They may physically have to prevent someone from hurting himself or herself or others; seek medical help; actively recruit networks of family, friends, or neighbors to stay with the helpee until the crisis has passed; arrange for some kind of immediate placement (for example, in a hospital or shelter); talk to people involved in the crisis situation to alleviate the cause of stress for the victim; or arrange for burial rites and insurance benefits. Crisis victims are often unable to take these actions themselves and need to feel they can depend on others for a short while.

The dependency of a crisis victim on the helper is accepted by the helper until such time as the victim is ready for a referral or to take over for himself or herself. The counselor may need to keep in contact and share information with others involved with the helpee in order to reach this stage of readiness and have other people provide support for the crisis victim.

Reactions to crises involving loss, such as divorce and death, have distinct stages, as noted by Krantzler (1973) and Kübler-Ross (1969). The initial reaction is one of shock and denial. The feeling is "this can happen to others, but not to *me*." Helpers provide empathic support during this stage. As the denial fades, anger emerges, and helpers often receive the brunt of this anger. Acceptance and then coping occur when the helpee can remobilize coping strengths and resources and begin to plan and implement action leading to recovery.

Another skill used by crisis workers in certain kinds of situations is confrontation, whereby the helper shows the helpee discrepancies in or ramifications of the crisis situation in order to stimulate immediate action. An example is a helper saying to a remorseful alcoholic who has beaten his wife while on a binge, "It's hard for you to control your temper when you drink. Now your wife is in the hospital, and she is thinking about pressing charges. In any case, she says she'll take the children and leave this time. It seems to me that you have to decide what you want to do about your drinking. If you want to keep your family, you'll have to deal with it. Let's go over the options you have." In a sense, this helper is telling the helpee to "shape up or ship out." Often this kind of "shock treatment" confrontation is necessary to get someone off dead center and moving in some direction.

In crisis situations, time is often a crucial variable; helpers do not have the luxury of building up long-term helping relationships before attempting problem solving. Therefore, confrontation often occurs earlier in crisis intervention than in other forms of helping. However, it is possible to be empathic and confrontational at the same time; the helper's tone of voice, body posture, and facial expression can make the difference between hostile and constructive confrontation. Constructive confrontation is not merely negative: it includes the acknowledgment of the helpee's strengths and ability to choose a course of action. When using confrontation, it is vital that you do so for the helpee's benefit, not to let off your own steam, prove your superiority, or impose your will. In other words, the confrontation can be assertive rather than aggressive, and can be in line with your helping objectives.

In addition to the knowledge required for specific types of crises (for example, drugs, alcohol, child abuse, suicide), human services workers involved in crisis intervention need to become thoroughly familiar with the sociological, economic, and cultural characteristics of the community in which they are working and the resources available within the community. Such knowledge is needed to make good referrals, especially when time is of the essence.

Following are three examples of crisis intervention.

Betty, a 19-year-old university student, came to see me unexpectedly. I had met her once, when she accompanied her roommate to my office; but I had had no direct contact with her. She was in obvious distress and incoherently blurted out that she had some pills and was seriously thinking about taking them because she "didn't want to live anymore." I immediately notified my secretary to hold all calls and appointments and spent the next couple of hours talking with her. She told me that her fiancé and brother had been killed in an automobile accident six weeks before and that she just couldn't get over it. She didn't feel she had anything to live for. She cried . . . I held her . . . we went over her loss and her anger at being left alone, over and over again. Finally, when she was completely exhausted, I asked her if she thought we could work together to find some meaning in her life. When she tearfully agreed, I pushed for some commitment: Would she give me the pills and promise not to do anything to herself until I could see her the next day at 3:30? I asked her to look me in the eye and promise that much. It took some time for her to be able to do that. I then suggested that it might be helpful for her to have some friends with her. She said that she had friends here, including her roommate, but that she had tried to keep her grief from them. I secured her permission to telephone her roommate and to arrange for constant companion-ship until our next appointment. This was arranged to everyone's satisfaction. The next day, Betty felt that the crisis was over and that we could begin to meet regularly to work her problem through without fear of suicide. Whether or not Betty's threats were legitimate is not important. I would never ignore such a threat nor discount it, and I do not leave clients until I have secured a commitment that they will not harm themselves.

Ari, 24, who worked at a beauty salon, phoned his employer to tell him that he was "sick" and could not come into work. He sounded very agitated and upset, and Robert, his boss and a hot line volunteer in his spare time, sensed that something was very wrong. Robert went over to Ari's house and found him disheveled, wild-eyed, and disoriented. Ari talked about dying and seemed inco-herent. Knowing that Ari's family lived in the Middle East and that war had broken out, Robert thought that was the cause for this upset and tried to calm Ari down by talking quietly. It turned out that the precipitating crisis for Ari had occurred at work the previous day when Ari had overheard other workers and some customers making racist comments about him. Ari had worked in this shop for ten years and was devastated by the abrupt loss of what he had experienced as safety and security. This had pierced his defense of keeping busy to avoid feeling and thinking about the war. Robert communicated empathy and demonstrated calm, supportive behavior. He encouraged Ari to return to the shop with him and he was able to soothe and reduce tensions among his staff by his manner and actions. After work that day, Robert took Ari to a community center where he could receive some group support from other Arab-Americans awaiting word about their families and friends in the Middle East. In this case, Robert's supportive communication skills as well as his active strategy of locating a resource to provide group support enabled him to help Ari to emerge from his terror and deal more effectively with his situation.

Mr. Budlong notified the personnel office that Tom, one of the mailroom clerks, had been arrested for drunk driving over the weekend. He complained that Tom's behavior had become quite erratic over the past month and that Tom, who had been the most dependable and conscientious mail clerk and was under consideration for a promotion, had "completely changed colors." Mr. Budlong was unsure what to do. Ms. Spitz, the personnel manager, phoned Tom at home (he was out on bail) and asked if she could come talk to him. She found him despondent, withdrawn, and concerned about the outrage of his parents as well as worried about losing his job. He mentioned that he had never been in trouble before and that he just didn't know what had happened or why. In talking to him, she found him confused and somewhat disoriented. After probing about changes that might have occurred in Tom's life the past few weeks, Ms. Spitz discovered that the pastor of Tom's church, who was a significant role model for Tom, had committed suicide four weeks earlier. Sensing that this was the cause of Tom's current difficulty, Ms. Spitz was able to engage Tom in discussion about his loss and to arrange a referral to a legal aid society and a community counseling center. In this case, an understanding of crisis theory and the judicious use of questioning enabled a caring, interested helper to effect referrals.

Exercise 8.7   Reenact the hot line role play in Exercise 8.6, using the following situation. A woman calls you on the hot line and hysterically says, "Hello, my 22-year-old son tried to kill me. He actually went at me with a knife. I can't believe it. He's acting like he's crazy. I don't know what to do. I'm calling from the corner phone. I'm scared to go home but I can't call the cops. He's my son. I love him. Help me. What should I do?" Compare your reactions to this exercise with your classmates' reactions, and discuss your choices of crisis intervention strategies.

Exercise 8.8   Which of the following statements do you believe to be true and which are myths? Compare your answers with other students' and discuss your differences and agreements.

1. Nice girls don't get raped.
2. It is better to give a robber what he wants rather than resist.
3. Divorce is always harmful to children.
4. It is not really rape if you did not resist.
5. People who lose their jobs are obviously partially responsible.
6. Strong people do not crack up in crises.
7. Alcoholics go on binges because people drive them to it.
8. A man has the right to rape a woman who dresses seductively or accepts rides.
9. People who have sudden financial reverses must have brought it on by poor planning.
10. People who threaten to commit suicide are not really likely to do it.

Exercise 8.9   Imagine that you are a female sophomore in a dorm. Your roommate of two years is a close friend; however, she has changed during the summer. Now she is sloppy, skips classes, drinks a lot, and either stays out all night or has a boyfriend in. You're uncomfortable with the fact that she has sex with her boyfriend in your room. In each of the following columns, list issues that you believe each person would have to deal with in relation to the other two people. In terms of dependency, time, attention, and compromises, what are the implications for each person? How do you think each of these people would handle anger, recrimination, and guilt?

Student	**Roommate**	**Boyfriend**

Exercise 8.10   In this exercise, you will role play the people described in Exercise 8.9 and the counselors assigned to each one. Each of you should role play all three counselors and determine in each case what the problems are, whose problems they are, who is responsible for the solutions, and what the prognosis is for successful resolution. What did you learn from considering the situation from three different perspectives?

☐ Summary

This chapter presented an overview of crisis theory, its relation to helping theory, and the practice of crisis intervention. The extension of crisis intervention theory to disaster and trauma was discussed. We formulated six classes of emotional crises, in order to differentiate between developmental transitions and situational traumas. Many people—nonprofessional, paraprofessional, and professional—deal with crises, either through face-to-face contact, over the telephone, or through various outreach programs.

Our review of crisis theory indicates that short-term intervention can be effective and that available support networks are critical components of crisis intervention. It is interesting to note that people in crisis can be very open to change; thus, crisis can lead to positive results, such as further counseling, recognition of personal strengths, and the ability to cope better with problems in the future.

Although crisis intervention sometimes demands a helping approach different from that of the counseling model developed in this book, it still requires good communication skills. The helping strategies discussed in Chapter 6 can be modified and applied to crisis intervention. In addition, we presented a synopsis of brief therapy as a viable form of crisis intervention because of its active focus on problem solving.

This chapter also included a review of the stages and steps of crisis intervention and the skills needed to practice it.

☐ References and Further Reading

Ashinger, P. (1985). Using social networks in counseling. *Journal of Counseling and Development, 63,* 519–521.

Ausubel, D. (1958). *Drug addiction: Physiological, psychological, and sociological aspects.* New York: Random House.

Budman, S. H., & Gurman, A. S. (1988). *Theory and practice of brief therapy.* New York: Guilford Press.

Caplan, G. (1961). *An approach to community mental health.* New York: Grune & Stratton.

Caplan, G. (1964). *Principles of preventive psychiatry.* New York: Basic Books.

Farber, M. (1968). *A theory of suicide.* New York: Funk & Wagnalls.

Farberow, N. L., & Schneidman, E. S. (Eds.). (1965). *The cry for help.* New York: McGraw-Hill.

Fiefel, H. (1959). *The meaning of death.* New York: McGraw-Hill.

Figley, C. R. (Ed.). (1985). *Trauma and its wake.* New York: Brunner/Mazel.

Fisch, R., Weakland, J., & Segal, L. (1982). *The tactics of change: Doing therapy briefly.* San Francisco: Jossey-Bass.

Gist, R., & Lubin, B. (Eds.). (1989). *Psychosocial aspects of disaster.* New York: Wiley.

Horowitz, M. J. (1986). *Stress response syndromes* (2nd ed.). Northvale, NJ: Jason Aronson.

Krantzler, M. (1973). *Creative divorce.* New York: Evans.

Kübler-Ross, E. (1969). *On death and dying.* New York: Macmillan.

Laurie, P. (1967). *Drugs.* Baltimore, MD: Penguin.

Lester, G., & Brockopp, G. W. (1976). *Crisis intervention and counseling by telephone.* Springfield, IL: Charles C Thomas.

Lester, G., & Lester, D. (1971). *Suicide.* Englewood Cliffs, NJ: Prentice-Hall.

Lieb, J., Lipsitch, I., & Slaby, A. (1973). *The crisis team: A handbook for the mental health professional.* New York: Harper & Row.

Lindemann, E. (1944). Symptomatology and management of acute grief. *American Journal of Psychiatry, 10,* 141–148.

Lystad, M. L. (Ed.). (1988). *Health response to mass emergencies.* New York: Brunner/Mazel.

Macdonald, J. M. (1975). *Rape: Offenders and their victims.* Springfield, IL: Charles C Thomas.

McGee, R. K. (1974). *Crisis intervention in the community.* Baltimore, MD: University Park Press.

Menninger, W. C. (1978). *Psychiatry in a troubled world.* New York: Macmillan.

Mills, P. (1977). *Rape intervention resource manual.* Springfield, IL: Charles C Thomas.

Okun, B., & Rappaport, L. (1980). *Working with families: An introduction to family therapy.* Pacific Grove, CA: Brooks/Cole.

Osterweis, M., Solomon, F., & Green, M. (Eds.). (1984). *Bereavement.* Washington, DC: National Academy Press.

Parad, H. J., & Parad, L. G. (Eds.). (1990). *Crisis intervention, book 2: The practitioner's sourcebook for brief therapy.* Milwaukee, WI: Family Service of America.

Richman, J. (1986). *Family therapy for suicidal people.* New York: Springer.

Slaikeu, K. A. (1984). *Crisis intervention: A handbook for practice and research.* Boston: Allyn & Bacon.

VandenBos, G. R., & Bryant, B. K. (Eds.). (1986). *Cataclysms, crises, and catastrophes: Psychology in action.* Washington, DC: American Psychological Association.

Watzlawick, P., Weakland, J., & Fisch, R. (1974). *Change: Principles, problem formation, and problem resolution.* New York: Norton.

Wright, K. M., Ursano, R. J., Bartone, P. T., & Ingraham, L. H. (1990). The shared experience of catastrophe: An expanded classification of the disaster community. *American Journal of Orthopsychiatry, 60,* 35–43.

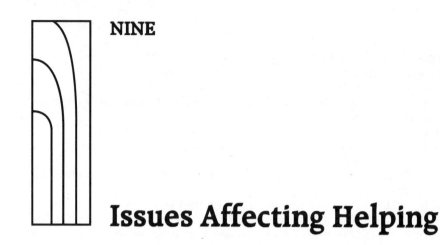

NINE

Issues Affecting Helping

Understanding the first two dimensions of the human relations counseling model, the stages and the strategies, will not, by itself, allow us to completely understand the helping process. Our personal values and points of view, represented by the model's third dimension, also affect the helping process, and unless we recognize this fact, our values and beliefs can seriously hamper our usefulness as helpers.

Most of us choose friends with whom we can relate easily and feel free to share our true feelings and beliefs. We enjoy being around people with similar value systems. Fortunately, we have the freedom to choose who will and will not be our friends. Helping relationships are different in that as helpers, we are not always able to choose who will come to us for help, to choose with whom we will work. If helpers are willing to work only with those who share their values, they should acknowledge this to prospective employers and employees, because they will be unable to develop effective helping relationships with people who do not share their values. Values, beliefs, and ideology can have positive or negative effects on helping relationships. It is necessary to explore this fact to increase our own self-awareness and become more aware of our impact on those with whom we work.

There are many societal, professional, and personal issues that affect helping relationships. It is not the intent of this chapter to explore all these issues in depth, but to make you aware of some of them. The consciousness-raising exercises and the reading list at the end of the chapter will enable you to study these issues further.

☐ Personal Values

Which personal values affect a helping relationship? Our attitudes and feelings about kinds of people, what is "good" or "bad," what is acceptable or unacceptable, what is important for choices, and what makes people tick lay the foundation for our value system. Our beliefs underlie our values; thus examining and reconsidering beliefs can help to clarify values.

The traditional models of helping maintained that if helping relationships were objective, distant, and neutral, the helper's values and beliefs would not contaminate the relationship. However, in recent years we have recognized that in any interpersonal relationship, whether a helping relationship or not, values are transmitted either directly or indirectly between the participants.

Thus, it is necessary for helpers to become aware of their value systems in order to aid helpees in clarifying their own. The point is that if you are aware of your own values you are less likely to impose them indirectly on others. Being aware of your own value system enables you to "lay it on the table" and declare your values as such. You also will become more aware of other people's value systems, which will help you understand and accept differences among your value system and others'. Discussion of values can enhance the helping relationship if both people are genuine and empathic. Trying to change another person's value system is fraught with dangers, but helpers can at least make clients aware of their own values in order to increase their self-understanding and ability to make effective choices.

Values Clarification

Values clarification is a systematic, seven-step approach developed by Simon, Howe, and Kirschenbaum (1972) that helps people process their values through structured exercises. The seven steps to values clarification they describe are as follows:

1. Prizing and cherishing one's beliefs and behaviors
2. Publicly affirming, when appropriate, one's beliefs and behaviors
3. Choosing one's beliefs and behaviors from alternatives
4. Choosing one's beliefs and behaviors from alternatives after consideration of consequences
5. Choosing one's beliefs and behaviors freely
6. Acting on one's beliefs
7. Acting on one's beliefs with a pattern, consistency, and repetition

The strategies developed by Simon, Howe, and Kirschenbaum are designed for use in groups and classrooms, but they can be modified and used in one-to-one situations. Discussion of values is a form of teaching and provides information and alternative viewpoints; it differs from the imposition of values, which insists that there is a "right" view to adopt.

Glaser and Kirschenbaum (1980) present examples of questions that helpers can use to aid helpees in the seven-step values clarification process.

1. Prizing and cherishing one's beliefs and behaviors
   a. "Is that something that is important to you?"
   b. "Are you proud of how you handled that?"

2. Publicly affirming, when appropriate, one's beliefs and behaviors
   a. "Is this something that you'd like to share with others?"
   b. "Who would you be willing to tell that to?"
3. Choosing one's beliefs and behaviors from alternatives
   a. "Have you considered any alternatives to that?"
   b. "How long did you look around before you decided?"
4. Choosing one's beliefs and behaviors from alternatives after consideration of consequences
   a. "What is the thing you like most about that idea?"
   b. "What would happen if everyone held your belief?"
5. Choosing one's beliefs and behaviors freely
   a. "Is that really your own choice?"
   b. "Where do you suppose you first got that idea?"
6. Acting on one's beliefs
   a. "Is that something you'd be willing to try?"
   b. "What would your next step be if you chose to follow that direction?"
7. Acting on one's beliefs with a pattern, consistency, and repetition
   a. "Is this typical of you?"
   b. "Will you do it again?"

Obviously, it is important for the helper to know when to question and when to listen, when to elicit expression and when to reflect back the feelings expressed by the helpee.

One's personal values are intertwined with one's beliefs about gender, family, money, politics, religion, work, race, authority, and culture as well as with one's personal taste and lifestyle. Values confusion usually results in interpersonal difficulties, the major reason that many people seek help from agencies and institutions. If helpers are uneasy about their values in any one of the previously mentioned areas, they may deliberately and subtly avoid those areas in helping relationships. For example, if you find that you never seem to get around to talking about sexuality with helpees who are involved in sexual relationships, you might ask yourself what's going on and why. You may find that you are the one who is uneasy about discussing sexuality, not the helpee.

Values Sharing

For helpers to create the necessary empathic conditions for effective helping relationships, they must be able to accept people with different value systems. This necessity can create quite a dilemma for you as a helper. How can you be both genuine *and* nonjudgmental with someone whose values you dislike? How can you acknowledge the dignity and worth of an individual whose actions, values, and ideology are unacceptable to you?

Accepting people means that you respect them as dignified, worthy individuals. It means that you recognize that they have as much integrity and right to be alive as any other human being. This does not mean you have to accept their behavior or concur with their values. Neither does it mean you cannot feel and express anger and disagreement when appropriate. It does mean that you can tolerate differences, ambiguity, and uncertainty, that you can accept that what is right or good for one person may not be right or good for another.

Responsive listening and sending "I" messages can help you communicate not only respect for the helpee as a worthwhile human being but also your genuine feelings and reactions along with your personal values and views, without imposing them on the helpee. In this way, the helper aids helpees in learning to acknowledge and evaluate their own values. By sharing our values with helpees, we may be adding options and alternative values for them to consider. By the same token, we may gain insight into our own value systems.

An example of values sharing occurred recently in my office when a couple came in for marriage counseling.

Counselor: Can you try to tell me what you see as the problems in this marriage?

Husband: Yes. She's not a good wife. She won't have sex with me and she's always going out with friends after work and she's not a good mother either—our boys don't listen to her. I make four times as much money as she—I give her a nice house, everything, and she won't even let me touch her.

Wife: You're always criticizing me, telling me what to do, and when I say no to the kids, you say yes. I'm fed up with this. I'm not your slave and I won't follow your orders. You make me sick.

Counselor: *(to husband)* It's important to you that your wife do things for you the way you want them to be done because then you would know she cares for you.

Husband: That's right! That's her job. Without me, she'd have nothing. Her job pays peanuts. I'm a self-made man; I work very hard and I give her everything.

Counselor: *(to husband)* You're angry and disappointed because you feel your wife is not meeting her obligations to you.

Wife: *(interrupting)* He always wants more and more. He's angry because finally I have my own friends and my own life. I don't care that I don't make much money. I like what I do and having some freedom now that the boys are in high school.

Counselor: *(to husband)* I'm having a hard time with your concept of a wife's obligations. However, I do see that, in your view, she is rocking the boat and that it's hard for you to understand or look at it any way other than as a sign that she is disloyal and uncaring.

In the preceding excerpt, I expressed my views without deriding or punishing the husband. This seemed to help both husband and wife to begin to acknowledge and evaluate some of their own values and beliefs about sex roles and marriage.

Another example of values sharing is provided by the following example of a 19-year-old college student who was talking about flunking out.

Client: My dad's going to be very angry about losing all this money. I'm going to see if I can get my tuition back.

Counselor: You're really afraid of your dad's reaction to your flunking out again.

Client: I'm going to tell them I was sick. I think I'll tell them I had an operation and that's why I couldn't attend classes. Don't you think they'll give me my money back? They did it for my girlfriend. She had her appendix out.

Counselor: It seems to you that you can ease your dad's reaction and anger at you if you can at least get some money back for him . . . the money will make it all right.

Client: Yes . . . that's what I'll do. Do you think they'll give me my money back? It's so much money . . . over a thousand dollars.

Counselor: I'm uncomfortable with claiming illness as the reason for not going to classes.

Client: Really? Why?

Counselor: To me, that would be lying. You must think I'm pretty square, but I want you to know how I feel.

Client: Um-m. It's always worked before.

Counselor: Sometimes the ends seem to justify the means for you, huh?

Client: M-mmm. I never thought of it like that.

The session then focused on the client's need to defend herself against her father's reactions. She became aware of the manipulative devices that had worked for her in the past. The counselor's statement of values did not inhibit the relationship. Two weeks later, the client reported that she had received a 75 percent rebate and that it had not been necessary for her to lie to get it.

Helpers who have strong ideologies are often dismayed by the philosophy that they should not impose their values on helpees. But imposition of values and beliefs rarely results in growth and independence. Rather, it results in submission to or withdrawal from the counselor. Helpers can share their ideologies and offer them as options for consideration, but it is important for all choices and decisions to be the helpee's, not the helper's.

An example of withdrawal occurred in a counseling session in which the counselor was working with a couple. The wife was concerned about having children because she did not want to give up her career; her husband felt it was important for children to be with their mothers full time for the first five years and that, since his job was more important than his wife's, she would have to be the full-time parent. This couple were in their 30s and because most physicians believe it becomes more dangerous to have children as middle age approaches, it was important that they resolve this issue before too many years passed. Thus, they went to a community agency for counseling. They were assigned to a female counselor who identified strongly with the wife and immediately proceeded to berate the husband for being a selfish male chauvinist. By so strongly interjecting her value system, this counselor had completely alienated both husband and wife by the end of the first session, and the couple chose not to return. The wife commented afterward that despite her anger at her husband, and even though she agreed with the counselor's values, she so resented her husband's being attacked that she became protective of him. If the counselor had used responsive listening and had spent time developing a relationship with the couple and drawing them out rather than reacting to their struggle, she would have been able to suggest some alternatives reflecting her values for their consideration. Thus, she would have had a better opportunity to expose them to different values and perhaps to help them.

Rogers (1967) suggests that if you continue to have negative feelings about a client you should first do some "homework" about those feelings, to determine where they are coming from and whether or not they relate to some of your own concerns. You can then check them out with the client by focusing on the areas causing the negative feelings, and if they continue to persist, you can finally confront the client, using "I" statements. Examples of this kind of confrontation are "I find myself feeling angry whenever I hear you talk about your husband in such a deprecating way, such as when you said that 'all men are really babies and Joe is no exception.' I'd like to talk about this with you because my angry feelings are getting in the way"; or "I'm having a hard time listening to you talk about spending money on fixing your TV when you know that Jamie needs to see a dentist. I know that for me a kid's health care is more important than TV. I don't want my values to get in the way, and I want to try to understand. Can we talk about this some more?"

There *are* going to be situations in which you find it difficult to be genuine and nonjudgmental at the same time. It would be irrational to believe you can like and work well with all people. If after you have tried talking it out with the helpee you are still unable to feel genuine positive regard for him or her as a person, you have certain options to consider: (1) locating another helper; (2) seeking help (from a supervisor or colleague) in working the problem through with the client; or (3) limiting your relationship to the accomplishment of specific, immediate, concrete goals, such as obtaining food stamps, processing papers, or providing factual information.

The following exercises will help you become more aware of how you react in uncomfortable situations.

Exercise 9.1   In small groups, role play the roles of helper and helpee in the following situations and then discuss your feelings and reactions to see if different people role play in similar or different manners. The purpose of this exercise is to help you become more aware of the effect of your values on your helping. Some controversial topics that will get you in touch with your values and others' are as follows:

1. A 12-year-old comes to see you because she is pregnant and wants to keep her baby.
2. A nurse's aide comes to see you because she is in conflict about the issue of euthanasia with regard to a terminally ill 37-year-old woman in great pain, who keeps pleading for an overdose of drugs.
3. A 16-year-old boy tells you he is having a homosexual affair with his 17-year-old cousin.
4. A couple is considering an interracial marriage.

Exercise 9.2   This exercise is anxiety provoking and should be used at the discretion of the trainer, who can assess how cohesive and trusting the group has become. In every group there are some people to whom we are immediately attracted and others whom we instinctively avoid. We do not usually check out these initial reactions. Look around the room and pick out the person with whom you would least want to be. Go over in your mind all your reasons and feelings as you continue to look at this person. Then go talk to this person and see if you can genuinely, empathically relate to him or her and share your feelings. See if you can get in touch with your avoidance reactions and with your own anxieties. How did you feel if you were picked? Process this in small groups.

Now let's do an exercise that will help you clarify some of your own values in relation to helping.

Exercise 9.3   For each question, rank your response: SD (strongly disagree), D (disagree), A (agree), SA (strongly agree). When you have completed the questions, go over all of your answers and look for a pattern in them. At the conclusion of the exercise, each member of the group can share something he or she has learned about himself or herself, without necessarily revealing all of his or her answers.

1. Which kind of person would be the most difficult for you to work with?
   a. a person 60–80 years old
   b. a person 40–60 years old
   c. a person 20–40 years old
   d. a person 15–20 years old
2. I would most rather work with
   a. a high school student
   b. a middle-aged person
   c. a college student
   d. an elderly person
3. It would be most difficult for me to work with
   a. a person of a different race
   b. a person of a different sexual orientation
   c. a person of an obviously different socioeconomic class

    d. a person with an obvious physical handicap

4. I would have difficulty working with someone who confessed to
    a. lying
    b. cheating
    c. promiscuous sexual relations
    d. heavy drug use

5. I am most comfortable working with people who are
    a. verbally articulate
    b. actively gesturing
    c. pensively attentive
    d. challengingly resistant

6. My overall helping style is probably
    a. fairly directive
    b. fairly nondirective
    c. somewhere in between directive and nondirective
    d. all of the above, depending on my mood

7. I am most comfortable with people who are
    a. independent and able to take responsibility for themselves
    b. helpless and needing my direction
    c. resistant to helping
    d. insistent on egalitarian collaboration

8. I am most comfortable talking about
    a. sexual matters
    b. money matters
    c. pain and loss
    d. death

9. I feel uncomfortable expressing feelings involving
    a. loving
    b. hating
    c. resentment
    d. failure

10. I consider myself to be
    a. self-controlled
    b. spontaneous
    c. forgiving
    d. unable to forget past slights

11. For me the most important of the following values is
    a. security
    b. freedom
    c. social recognition
    d. affiliation

12. Which of the following statements most closely fits your own beliefs?
    a. Cleanliness is next to godliness.
    b. Honor thy father and thy mother.

    c. A stitch in time saves nine.

    d. Do unto others as you would have them do unto you.

13. For me the most important of the following values is
    a. happiness
    b. inner harmony
    c. mature love
    d. success

14. It is difficult for me to
    a. let things take their own course
    b. take action and responsibility for getting what I think is necessary
    c. let someone else take the initiative
    d. share equal responsibility with others

15. I believe
    a. there is one universal set of moral values
    b. moral values are individual
    c. moral values should be taught outside the family
    d. helpers should teach moral values

Sexism

In the past 20 years, changes in the economic, social, and legal position of women have resulted in dramatic shifts in the traditional values, expectations, and life goals of both men and women. These changes are reflected in our educational and child-rearing practices as well as in different types of family structures. Today's generation of adolescents and young adults is making very different career, family, and lifestyle choices from those of their parents and grandparents. For example, many more women are not only entering but are also advancing in professions and occupations that were heretofore the domain of men (such as law, computers, finance, government leadership, aviation, medicine, management, high technology, the military). There are increasing numbers of dual-career couples (some even commute), single men and women choosing to have a child naturally or by adoption, single parents by divorce or widowhood, same-gender couples, and men and women single by choice, all of whom struggle to balance individual, career, and family needs and demands.

    Today, there is more acceptance and understanding of gender differences due to biological and socialization variables, and there is a continuous demand for equal treatment of and power for both genders. The work of Gilligan (1982) and Miller (1976) has been crucial for understanding that women are socialized to be more relational and attached than men, who are socialized to be more autonomous and separate. Throughout the life span, women deal with issues and transitions from a nondominant perspective in male-dominant culture. These different developmental paths have profound impacts on individual, couple, and family experiences.

Attention to gender inequities and differences in psychological development necessitates new considerations in helping theories and the helping process. Feminist therapy approaches address these issues, but *all* helpers need to consider nonstereotypic career development, the gender issues of life-span development, and more complex family and couples counseling as people deal with dual-career pressures, divorce, widowhood, singlehood, homosexual relationships, and single parenting. Equally important is attention to counseling for such issues as unplanned pregnancy, abortion, adoption, infertility, rape, sexual abuse and harassment, and decisions about sexual orientation.

Despite the progress that has been made in the past decades, stereotypes have not yet been erased. Regional, generational, and ethnic variables affect the degree and pervasiveness of sexism in our personal and professional lives. Studies still indicate a widespread occurrence of sexist counseling. In fact, the danger is more insidious than before in that many helpers believe that they are open-minded and *not* sexist, while their actual helping reflects basic stereotypic beliefs and values.

A classic study by Broverman and his colleagues (1970), clinical psychologists known for their research into sex roles, indicated that mental health workers had different standards of mental health for men and women. The research found that many mental health workers perceived males as "healthier" than females in general and "healthy" women as more submissive, less independent, less adventurous, less competitive, more easily influenced, less aggressive, more excitable in minor crises, and more emotional than "healthy" men. A study by O'Malley and Richardson (1985) has updated the Broverman study, finding that while counselors' standards are a reflection of sex stereotyping in our society as a whole, they now perceive both typically feminine characteristics and typically masculine characteristics as appropriate for the normal adult. Thus, more recent attitudes accept that healthy adults possess both masculine and feminine characteristics— although, presumably, a man should still have masculine characteristics and not strong feminine ones, and a woman should have feminine characteristics and not strong masculine ones.

Whitely (1979) and Smith (1980) reviewed the research literature and concluded that present evidence has not demonstrated counselors or the therapy process in general to be sex-biased. On the other hand, Sheridan (1982) questions those findings, concluding from her review of sex bias in therapy research that the literature on gender bias in therapy is a study in discrepancy due to differing perspectives. Lichtenberg and Heck (1983) point out the discrepancy between the evidence and our beliefs, between the processes of believing and knowing. It is possible, they assert, "that our awareness of and sensitivity to the gender issues have reached the point where sex bias in counseling can no longer be demonstrated" (p. 104).

It is not yet clear what impact the sex of the helper and the helpee has on the helping process. One investigator (Parker, 1967) found that male therapists tended to be more nondirective with females than with males, while another (Heilbrun, 1970) found that among helpees leaving counseling prematurely, dependent females showed a preference for more directive therapist responses than males did. With regard to same-gender and opposite-gender relationships, West (1984), studying physicians, found that female helpers are interrupted and challenged by both male and female helpees more frequently than are male helpers. Rubin (1985) confirms the effect of gender differences in helping relationships by describing the higher level of affective self-disclosure (feelings) that male clients report experiencing with female therapists than with their former male therapists. She believes that gender differences in attitude and relationships do exist and that they can influence the content and process of the helping relationship. Likewise, male-female helping relationships may contain higher levels of sexual attraction variables than same-gender helping relationships. The point is that there are most likely gender-linked differences in helping relationships; however, there is no evidence that same- or opposite-gender helping relationships are better or worse than each other—merely different.

Sexual harassment and abuse by the helper is a serious concern of the mental health profession. While the number of complaints has risen in recent years, it is thought to be vastly underreported (Gartrell, Herman, Olarte, Feldstein, & Localio, 1987). There is no equivocation by the mental health profession that such abuse is unethical and harmful. While some states have legislation making this action a civil and/or criminal infraction, unless the helpee is willing to bring charges against the helper who violates this ethical code, little can be done formally. Because of the increasing number of complaints, professional associations and state legislatures are paying more attention to this issue, hoping to have an impact on the incidence rate.

Sexist counseling occurs when the helper either overtly or covertly uses his or her own sex-role ideology as a basis for helping. For example, when a high school guidance counselor discourages a young lady from taking an electronics course or considering a career in carpentry because these are "male occupations," it is sexist counseling. Discouragement may be subtle ("Have you really thought about how difficult it will be working with only men?") or less subtle ("Do you really want to risk your femininity this way?"). Often helpers are not even aware that they are discouraging helpees in this manner.

Another example of sexist counseling is the helper who suggests to a woman that she shouldn't think about working while her children are young because "mothers of young children should really be home full-time." Sexist counseling also occurs when boys are discouraged from becoming teachers or nurses because it is "women's work," from taking home economics or child-care courses, from expressing emotions and feelings, or when they are chastised for not being athletic. Perhaps a more insidious example is the helper who suggests that a woman is not a "true feminist" if she enjoys staying home with her family and engaging in homemaking, and who urges her to seek a career to "fulfill" herself.

In addition to the previously mentioned kinds of sexist counseling, the American Psychological Association Task Force on Sex Bias and Sex-Role Stereotyping in Psychotherapeutic Practice (1978) reported sexist counseling practices such as fostering prolonged dependency relationships; having sexual contacts with clients; encouraging women to submit to oppressive and stereotyped sex-role behaviors within marital, sexual, career, and employment situations; blaming the woman who is a victim of assault or of sexist practices by others; interpreting women's problems in terms of sexist theoretical concepts; providing a spouse with unauthorized information about the client/wife; and failing to use appropriate community referral resources.

Sexism exists in most organizations and is often subtle and difficult to identify. Aside from obvious sexist practices such as paying different salaries for the same job functions, discriminatory promotion practices, and so forth, in many so-called progressive organizations women are assigned to leadership tasks involving human relations because "women are better at that kind of stuff." Women are also subtly discouraged from applying for positions supervising men and from behaving assertively in staff meetings and at conferences.

We now recognize the impact of sexism on men, as well; they are often discouraged from entering traditionally feminine occupations or looked down upon if they choose to assume primary parenting or housekeeping responsibilities. Nonsexist occupational, family, and lifestyle choices are just as necessary for men as for women. Although it is possible that sex-role stereotyping may be changing in our society due to affirmative action programs, the women's movement, and the economic necessity for more and more women to enter the labor market, it is important for helpers to deal with their own biases, the biases of others in their work settings, and the biases present in testing and informational materials in order to expand, rather than restrict, options for both males and females.

If you want to combat sexism actively in organizations and in human relations, you can promote day-care facilities for women in your setting; help promote flexible hours and jobs for women with families; actively agitate for the opening of jobs, courses, programs, and activities for both males and females in your work setting and community; and help reduce sex discrimination in recruitment and hiring as well as in training and placement. In other words, you can put into practice what you preach. Also, whenever possible, you can initiate group discussion among males and females within your organization and community about current sex-role ideologies and possibilities for expansion into personal, family, and work lifestyles. These activities will have a ripple effect in terms of your actual helping relationships, in that greater awareness results in more open, fairer helping relationships.

The following exercises will help you discuss sex-role ideologies within your training group.

Exercise 9.4   Divide your group into females and males. First, each individual ranks males and females in general from 1 (low) to 5 (high) on the characteristics in the following list. After each person has completed his or her rating, males and females compile group ratings, which they post on the board or on a large piece of paper, using the following form:

Male Group					**Female Group**	
Males	Females				Males	Females

1. serious intellectual pursuit
2. aggressiveness
3. emotionality
4. physical strength
5. nurturance
6. humor

Compare the results and discuss them as a large group. Are there greater differences *between* the male and female groups than between the ratings of males and females *within* each group? How do you explain these differences? What helped or hindered reaching a consensus in each group? Does the consensus agree closely with your individual scores?

Exercise 9.5   First decide how you feel about each of the following statements. Then, in small mixed groups, discuss them as freely as possible, remembering to practice responsive listening, to listen closely to the views of others, and to express your own values without imposing them on others.

1. Most women are capable of performing well as both worker and home-maker.
2. Women are not capable of becoming policemen or firemen.
3. Women should not make more money than their husbands.
4. Women with children under 5 years old should not work.
5. There is something wrong with people who are not married by the age of 35.
6. No man wants to work for a female boss.
7. Women are better suited to parenting than men.
8. Men have more leadership capabilities than women.
9. Men and women cannot have close relationships outside of a sexual relationship.
10. Women who are assertive and successful in their careers lose their femininity and their attractiveness to men.

Racism and Ethnocentrism

Like sexism, racism exists in our society; racist counseling exists when helpers allow their biases about different ethnic groups to contaminate helping relationships. Obvious examples of racist counseling in schools are when minority children are discouraged from enrolling in college-preparatory programs, from white-collar career exploration and post–secondary school education, and from joining school clubs and seeking office. The same discrimination and racist counseling occurs in work settings of all varieties, where minority groups are denied access to certain jobs or where tokenism is practiced. Racist counseling also surfaces in family service agencies when white middle-class helpers act on false assumptions about minority groups' family roles.

Racism can be combatted only with continual consciousness raising, study, exposure to different cultural value systems, and training of more minority helpers. It is important that helpers support equality for people of all races and that a concerted effort be made to train minority people for the helping professions so that we can all learn together to deal with our racist biases and to reduce racist helping.

The issue of how effective white helpers can be with minority helpees or how effective minority helpers can be with white helpees has been raised over and over again during recent years. I personally have seen many examples of minority counselors effectively helping white clients and vice versa. It is my belief that a skilled, competent helper who is conscious of his or her biases and values can work effectively with a wide variety of clients, regardless of race. However, conclusive research findings about this issue do not exist. I base my view on the belief that all people have the same kinds of psychological needs, problems, and physical sensations despite the obvious differences in circumstances and societal opportunities. Helpers' skills and specialized knowledge are more important than helpers' color, race, religion, or sex.

It is important for helpers to seek out nonracist information and tests, to provide for group interaction between people of different races, and to attempt to expand horizons for all people. Likewise, it is useful for helpers at all levels to keep up with the current literature and research regarding racism and to continue to expand their consciousness about different races.

Horror stories illustrating racist counseling abound. One Mexican-American student recently reported her dismay when her high school counselor told her she should not attempt to go to college because her father disapproved and it was "important for her to stick to her heritage." A very talented black graduate student was told "kindly" by his college counselor that he was "not appreciative enough of how far 'your people' have come and you want too much too soon . . . it takes lots of time and patience."

Many white helpers report that they feel uneasy when working with minority clients, fearing they will be considered racist if they confront or direct in any way. Black and white helpers who work with minority clients recommend that helpers bring up the issue of race during the initiation/entry stage of helping. An example might be "You're probably wondering how I, a white person, can possibly understand you and your experience as a black person." If the helpee is concerned about the race of the helper, then it can be discussed and explored at the outset. If the client is surprised by the introduction of the racial issue and states that it is not an issue for him or her, the helper can proceed to the concerns presented by the helpee.

A more subtle form of racism exists when helpers apply different strategies with clients from different ethnic groups, or when they modify the helping process in any way based on unsubstantiated attitudes and assumptions about race rather than on the nature of the helpee's problem. Some helpers have been heard to say that responsive listening is not effective with black helpees, that directive, behavioral techniques are more effective; however, responsive listening has been found to be equally effective in developing helping relationships with all helpees who possess normal verbal skills.

As international boundaries continue to diminish, we will be interacting with more cultures than ever before. As helpers, we need to become aware of our own values, acquire knowledge of whatever cultural group we are working with, and learn the helping strategies and approaches applicable to that group. Pedersen (1988) suggests that helpers need to be eclectic so as to deliver services appropriate to a variety of cultural groups. He urges helpers to figure out which of their basic values are shared by everyone, regardless of culture, and how cultures differ along a continuum. For example, styles of interpersonal communication range from direct to indirect, expectations of authoritarian figures range from powerful to weak, and degrees of individual control and responsibility range from high to low. D. W. Sue (1981) and S. Sue (1988) also stress how important it is for helpers to understand the helpee's view of the social environment so as to evaluate problems from the helpee's perspective. This will help in understanding cultural attitudes toward helping relationships and the helping process.

Obviously, further research is needed to define helping factors in different cultural groups, whether we are talking about international cultures or subcultures within our own society. The main points to remember are that you bring your own cultural values and perspectives to the helping process and that there are many differences in values, attitudes, and beliefs both among groups and among individuals within groups. It is also important to consider the different effects of gender, class, and assimilation experiences and opportunities within diverse cultures.

Thus our theories, techniques, and the profession itself are cultural phenomena reflecting our culture's history, beliefs, and values. Our norms are not necessarily applicable to another cultural group and we need to continually reassess our assumptions.

Exercise 9.6   This is a much-used exercise, with many variations, that helps you recognize discrimination. Divide into two groups—they may or may not be the same size. The class leader arbitrarily assigns one group to be the "haves" and the other group to be the "have-nots." The have-nots use armbands or some other means to distinguish them from the haves. For the next 30 minutes, the haves are to practice every form of discrimination they can conceive of such as placing all the have-nots together in a corner of the room, eating and having fun in front of them without letting them participate, taunting them, whispering about them, and so forth. At the end of the 30 minutes, without any processing, switch the group roles for an additional 30 minutes. After this, process your experiences: How did you feel as a have and as a have-not? What was easy and difficult for you? What kinds of leadership emerged in each of the groups? Whom did you find yourself allying with and staying away from? What did you do or not do?

Exercise 9.7   In small groups, see if you can agree upon at least five basic values that apply to all human beings regardless of racial and cultural differences. Then share your findings in the large group. After group discussion, rank each value SD (strongly disagree), D (disagree), A (agree), or SA (strongly agree). What did you learn from this exercise? Did you change your mind about any of the values listed?

Ageism

Ageism—perhaps a less obvious "ism"—is defined as imposing our own beliefs and values about what can or should be done at different ages onto other people. We know that age discrimination exists in the labor market: some people are considered too old for a job and some too young. Age discrimination also exists in human services, in that some helpers believe that older people cannot really be helped, and therefore avoid them. (Ageism usually refers to the problems of being elderly rather than to the problems of being too young.)

Helpers should try to understand that there are wide varieties and differences in individual development and that restricting opportunities by age may not be valid. Likewise, each developmental stage or age has something unique to contribute to our society, and we should begin to think more about the positive aspects of old age than about the problematic aspects.

The number of healthy, actively functioning people over the age of 65 has been steadily increasing in the past decades. From them we are revising our negative, restrictive expectations and attitudes about the health, emotional, family, and social issues of late adulthood. The meanings of life and possibilities for gratification in later years are limitless. Yet, we have many people over 65 who are poor and face health , housing, and economic problems, and we have few social supports and resources with which to help them. This population will be a target population for the helpers of the coming decades.

Helpers need to study the age-linked changes that occur in sensory, cognitive, motor, and affective areas, and the effects of environmental, social, and health factors on older people. For example, older people often feel inadequate because they can no longer financially support themselves. We as helpers can let them know that their poverty is not of their making and that we understand that their problems often result from inappropriate societal opportunities.

Time, patience, and empathic communication skills are essential to provide effective helping relationships for the elderly. In addition, reminiscing is an important strategy. The older person's strength is in his or her past. In contrast to younger people, the elderly are really quite secure in their pasts. Having already lived, they needn't worry about ruining their lives or dying while young. However, elderly people may occasionally regress and become upset about difficulties that occurred many years before. Reviewing with the elderly all they have done with their lives and focusing on the positive (as in crisis intervention strategies) can be helpful.

Advocacy seems to be one of the best strategies for working with the elderly. Helpers need to learn about the services and resources for the aged within their communities, to identify key people within helping organizations, and to familiarize themselves with bureaucratic procedures (for example, Medicaid and Social Security practices).

Again, we need to become conscious of our own attitudes and beliefs, to examine our stereotypic assumptions about different age groups. An example of ageist helping occurred when an 86-year-old man informed the social worker in the home for the aged that he wished to marry his 83-year-old lady friend down the hall. They were wondering if they could obtain a double room. The social worker reacted with horror and told him that he was behaving immorally, like a "foolish child," that "everyone knows people your age don't sleep together!" Another example of ageism is something that happened to me. When I applied for admission to a doctoral program, I was told that I would have to complete my studies before the age of 35 in order to have enough time to make a worthwhile contribution and justify my education. It is important for the helper to remember that although age may be important in making a particular decision, many other factors are also important.

Helpers can educate families and members of the community about the facts of old age and options for dealing with aged relatives. It is hoped that such education will lead to the reduction of discrimination and segregation due to age in schools and work settings as well as in facilities for the aged, and to increased incorporation of the aged into the mainstream of our society. For example, a group of volunteer workers in a neighborhood day-care center recently arranged for senior citizens to participate in activities with the children. This has proven to be so successful for both the children and the senior citizens that plans are under way for expanding this involvement of the elderly in the care of the young. (See Okun, 1984, for further discussion of these issues.)

Other Values Issues

Other values issues that may affect the counseling process relate to economic and social status, dress and physical appearance, language, religion, and health status. Helpers should deal with these issues as with the ones discussed in the preceding pages: by clarifying and expressing their values in a nonjudgmental manner and by collecting valid new information on the topics.

Keep in mind that the 1970s, with their social and legislative attention to issues of discrimination, have passed. The 1980s brought fiercer competition for more limited funding and a backlash reaction to social programs that was even more discriminatory than were pre-1970 attitudes. Examples include the hysteria about allowing children with AIDS to attend school and the widespread anger over the costs of providing special building access and education programs for the physically handicapped.

In the 1990s, the hysteria continues. As we struggle with even more stringent economic limits, prioritizing our social issues has become more complex and complicated. Current values issues include the right to life; the right to die; the civil rights of prisoners, homosexuals, and AIDS patients; birth technology; and organ transplants. In such an atmosphere, helpers need to be more socially aware and more flexible than ever before to make the most of the limited resources available to them and their clients.

☐ Ethical Considerations

Professional organizations such as the American Association for Counseling and Development (which includes the American Mental Health Counselors Association and the American Rehabilitation Counseling Association), the American Psychological Association, and the National Association of Social Workers have recently revised their codes of ethics to protect helpers from the public and vice versa. Most ethical codes are based on five fundamental ethical principles: (1) respect autonomy, (2) do no harm, (3) benefit others, (4) be fair, and (5) be faithful (Kitchener, 1988).

Ethical problems are complex, and it is impossible for helpers to avoid them. Those whose behavior is consistent with their definition of helping and who are committed to questioning their own behaviors and motives and seeking consultation from others are less likely to function unethically than those who are closed to such reflections. The ethical issues that appear to be particularly relevant in the 1990s are (1) privileged communication and sharing of confidential information, (2) conflict of interests, (3) record keeping, (4) use of tests and computerized programs, (5) dual relationships, and (6) misrepresentation.

Privileged Communication and Confidentiality

Most helpers do not have privileged communication, which means that they can be called on to testify in a court of law about the nature of their discussions with helpees and that their records can be subpoenaed. In some states certified counselors and certified psychologists do have privileged communication, but laws relating to this issue are not uniformly applied.

There are two kinds of confidentiality: legal and personal. The former depends on the laws of your state. The latter is of your own making, regardless of the policies of your work setting. In other words, if because of your setting (such as in a school or correctional institution) you are unable to treat information obtained in a helping session confidentially, you *must* make this clear to the helpee in advance of his or her sharing concerns with you. Then the helpee can choose whether or not to tell you anything that is controversial. If you are able to promise personal confidentiality, you must maintain it; under no circumstances should you reveal to anyone what has been told to you in confidence.

Recent court cases indicate that professional helpers can be held accountable if they do not warn potential victims of violence. Since each state has different legislation, mental health professionals encounter legal as well as ethical dilemmas in their attempts to determine whether a client is likely to commit psychological or physical harm to another person. Paraprofessional and nonprofessional helpers are also affected by this dilemma: should they report legitimate concerns about the welfare of a helpee to another person, whether it be a family member, a helping professional, or an official? For example, if they know that someone is contemplating suicide, should they tell someone? Who? Under what circumstances?

Conflict of Interests

There are times when you may experience a conflict of interests between your obligations to your organization and your obligations to your helpee. There are no blanket solutions for such cases. Each person must find an answer that can be lived with. For example, a school counselor recently related hearing from one of his clients the names of the students who had participated in a serious act of vandalism during the previous week at school. The counselor was torn between his desire to protect his client and his desire to deal with the offenders in a helpful manner. After several difficult sessions with his client, he was able to persuade the boy to tell his friends that he had told the counselor about the vandalism and to bring his friends with him to the counselor's office.

In another case, a street worker was told by a youth that his gang was going to break into a jewelry store that night. The street worker, after much soul-searching, called the youth back and told him that he was going to have to warn the owner of the jewelry store that a gang was planning a burglary. The counselor explained to the youth that he was worried about the physical safety of the owner of the jewelry store. He warned the store owner and, to his surprise, the youth kept returning and sharing confidences with him, indicating that perhaps he had hoped for this outcome.

It is important for helpers to remember that their primary responsibility during the helping process is to the helpee, not to any other individual or group. Thus, if a conflict of interests arises, helpers must make sure that they do not breach a helpee's confidentiality because of their own ignorance, insecurity, or ineptness, or for the good of some organization or group. The only justification for a breach of confidentiality is that the welfare of the helpee or some other human being is at stake.

Conflict of interests may also pose ethical dilemmas for workers in settings concerned with census count. One mental health worker in a psychiatric hospital reported discomfort when the director of the hospital suggested to the staff that they "treat all the patients as if they were at home as it's important to keep them here until their insurance runs out." Another mental health worker decided to leave his HMO employer because he was offered a financial bonus for every case he was able to counsel in fewer than five sessions. These types of stories abound in today's cost-conscious climate and are of concern to all of the professional associations.

Record Keeping

With regard to records, you may find it a good idea to write down only objective, behavioral information and to exclude subjective material such as your values, attitudes, and, above all, interpretations. Some helpers keep certain personal information and coded personal notes in their office files. If tape recordings are used, it should be with the full knowledge of the helpee, a full explanation of the intentions and purposes of the recordings, and the helper's guarantee that the tapes will be destroyed after they fulfill their purposes. If third-party (for example, insurance) payments are involved, discuss what information is provided to that party.

Testing

Some human services workers are involved in testing helpees. They may employ interest, aptitude, or achievement tests or a paper-and-pencil personality inventory that could be used for screening, placement, or some other kind of classification. It is ethical for helpers to administer tests only if they have had sufficient training and supervision in the administration of the particular test. In addition, helpers are ethically bound to determine and explain clearly to helpees the rationale for and purposes of the testing process.

A related issue concerns the use of test data—who receives these data and what use will be made of them. Does the helpee receive feedback in order to increase self-understanding? Is the helper able to interpret test data accurately and explain them to the helpee along with the meaning and validity of the test? In these cases, each helper must determine the extent of his or her ethical responsibility.

Remember not to discuss helpees with anyone in their family or with another staff member of your organization, such as a teacher, supervisor, or employer, without the helpee's specific verbal permission. This precaution is necessary to ensure trust and to communicate helper interest in helpee welfare.

Dual Role Relationships

A recent troublesome concern of ethical committees has been the potential harm of dual relationships between professional helpers and students, supervisees, clients, employees, or research subjects. Different roles involve different expectations of behavior, power, and obligations. It is likely that there will be conflict between these role expectations if the helper engages in any role other than that of the professional helper. Examples of dual relationships are forming a social or sexual relationship with a helpee, becoming a student's or supervisee's therapist, or entering into a professional helping relationship with a friend or family member.

Clear boundaries between the helper and the helpee, and clarification of the nature and process of the helping relationship, will mitigate against possible abuses of power. For example, if professional helpers conduct counseling or therapy with their students or supervisees, confidentiality may be compromised and the students' or supervisees' autonomy may be impaired. Furthermore, the helping process is nonevaluative, and teachers and supervisors are usually required to evaluate their students and supervisees, so objectivity will be damaged. The point is that dual relationships have a high potential for exploiting the helpee, who is in a less powerful position than the helper regardless of the circumstances. If this power differential is not acknowledged and both parties consent to another kind of relationship, the damage may be more insidious.

Misrepresentation

This can occur when a helper directly claims or indirectly infers knowledge, training, experience, and/or expertise with a particular type of client or a particular type of problem. This could be an issue for nonprofessional, paraprofessional, or professional helpers who are uncomfortable or unable to acknowledge their personal and helping limitations. Helpees are often too confused and eager for assistance when they seek help to check the qualifications and experience of the helper. Professional codes of ethics specifically require helpers to acknowledge their limitations. For example, a colleague came to me to discuss her dilemma when it became evident that one of her clients had multiple personalities. My colleague had never experienced this kind of case. She discussed her limitations with the client and offered her three options: (1) continue their work together, given the effectiveness of their helping relationship, with supervision and consultation from an outside expert; (2) referral to an expert for direct treatment; or (3) co-therapy with an expert. On the other hand, there are situations where a helper has not sought supervision or consultation, allowing the helpee to assume a level of experience and expertise above and beyond the fact.

No code of ethics can cover all situations and all circumstances. Ethical issues involving the use of computers are now being addressed by professional associations. How do we guard against the release of information from computers? How do we maintain a humanistic, socially conscious perspective as we become more reliant on high technology in the helping process? How do we deal with the fear of litigation? Appendix B contains a sample policy statement that I use with my clients. You will have to decide for yourself what course of action is right for you.

☐ Other Issues That Affect the Counseling Process

We will just briefly touch on some of the other issues that frequently arise in helping relationships. The readings at the end of the chapter will provide more in-depth coverage of these areas.

Reluctant Clients

A frequent concern of helpers in certain institutions centers on the reluctant client, the individual who is told he or she *must* seek help. This may cause an ethical dilemma for helpers who believe in the self-determination of helpees. After all, if people have the right to determine for themselves whether or not they want to be helped or "get better," how can we force our services on them? If you feel this way, you can just tell the helpee, communicating empathy and concern for his or her position while explaining your own.

Confrontation of resistance is more likely to engage reluctant clients than evasion. Sometimes the reluctant client is testing you, waiting to see how committed you are. And he or she may have all the time in the world to wait.

Self-Disclosure

Self-disclosure, the technique whereby helpers talk about themselves with helpees, is another controversial issue. At one time it was considered disastrous and unethical, but now many helpers believe that judicious self-disclosure can enhance the helping relationship and aid in problem solving. The guiding principle is that the helper's self-disclosure should be for the helpee's benefit. This means the helper does not burden the client with his or her problems but instead regulates the quality and timing of self-disclosure to help the client focus more on his or her concerns and to encourage exploration and understanding.

If the helper's self-disclosure is used to further the helpee's exploration and self-understanding, it will ideally allow the helper to become a positive role model, because it both reveals the helper's humanness and provides a look at the possibility of coping effectively with whatever the issue is. It will neither distract the helpee from pressing concerns nor adversely affect the relationship by showing up the helper's flaws. Rather, it will aid the helpee in focusing on the central issues of the moment and teach him or her that although all human beings have problems and can make mistakes, one can still learn from those mistakes and work through those problems.

An example of self-disclosure occurred recently in a pregnancy advisory service, where the volunteer helper was discussing the possibility of abortion with a 19-year-old single woman. The helpee was hesitant about considering abortion, but she felt trapped and unsure about having a child. When she began to contemplate the guilt feelings that might result from an abortion, the helper shared her own painful, guilty feelings from a similar experience and explained that there would most likely be guilt and discomfort resulting from any of the options being considered. The helpee seemed relieved to be talking to someone who had experienced those feelings and lived through them.

It is important that helpers have the capacity to share themselves intimately with other people, although they certainly must carefully choose the people they share themselves with and the appropriate time to do it. We expect helpees to share themselves, and we need to experience and understand that same process of sharing and relating. The underlying principle here is that we do not ask others to do what we cannot or would not do.

Change-Agentry

When helpers take active roles in bringing about political and institutional changes to combat discrimination and inequality, they are acting as change agents, or advocates. (The sections on sexism and racism in this chapter gave some examples of change-agentry/advocacy.) Increasing numbers of people in the helping fields are taking active stands against racism, sexism, and ageism; against poverty and other forms of inequity; and against discrimination toward those politically and organizationally involved and active. The rationale for this increased activity is that consistency between what one practices and what one preaches is necessary in order for changes in systems to occur and in order for effective helping relationships to occur, given that people are becoming more cognizant of the role that social influences play in psychological discomfort.

As helpers, we are becoming more and more aware of the problems that systems and institutions impose on helpees—problems that subsequently become labeled as the helpees'. As helpers become frustrated struggling against the limitations that institutions impose on them and on helpees, they may become more politicized and radicalized. This does not mean that they revolutionize or seek the destruction of the institution, but that they try to foster systemic, innovative change within the institution.

Issues that might necessitate change-agentry strategies are program control and budgets or the political process involved in staffings, organizational decisions, and public policies. Techniques of change-agentry include generating public support by disseminating written materials; organizing support groups; training staff; providing public education; mobilizing local, state, and national election task forces; pressuring elected officials; writing legislators; and participating in confrontations such as boycotts, demonstrations, and strikes when they are part of a total strategy or campaign.

Change-agentry requires planning strategies, determining goals, and designing a structure to reflect those goals, as well as making tentative implementation and evaluation plans. Although the change agents or innovators may assume initial leadership, one of the frequent priorities in institutional change is a more broadly based sharing of power among all members of the institution. If staff recruitment and training in human services institutions can result in the availability of facilitative helpers rather than authoritarian "experts," then the needs of the helpees will assume a higher priority than the needs of the helpers and the needs of the organization to maintain itself.

As an example, let's consider a federally funded mental health agency that experienced a great deal of chaos a few years ago. This chaos was reflected in high staff turnover, a large number of written and verbal client complaints and no-shows for appointments, and resistance to the agency from other community groups. The feelings of the staff were that the agency was not being run to meet the needs of the community it was supposed to serve. Two staff helpers became so frustrated with the bureaucratic red tape they encountered every time they brought up the idea of change that they finally recruited active support from members of the clerical and volunteer staff as well as from members of the community. As a task force, they prepared a written report of what they perceived to be the problems of the community and of the agency and the conflicting directives and policies, and presented it to their director and the senior staff. The report included a written proposal for a pilot project that would change the staff loads and include members of the community and some representative helpees at regular agency meetings.

Because their proposal was fully documented and specific, the director agreed to a two-month trial. At the end of the trial period, modifications were suggested, and some changes in leadership occurred when some senior staff members realized that they would be unable to muster the support necessary to revert to the previous way of operating. The new leaders encouraged members of the community to participate in agency policymaking and decision making and were successful in effecting many of the changes recommended in the original proposal.

Dedicated helpers are increasingly aware that communities must have a hand in the development of their own institutions and that those institutions must become more responsive to the needs of their client population. This means that services must be adequately provided to meet the unique needs of a particular community, not to provide research data for an erudite report or to provide training for paraprofessional and professional helpers from other communities. You must decide for yourself, as you consider your helping role and your own personal and professional ethics, just how involved you will become in system-wide changes.

The following exercises are designed to help you become more familiar with the issues just discussed.

Exercise 9.8   This exercise should help you experience the feelings of both a reluctant client and the helper of a reluctant client. In pairs, assign roles of helper and helpee. Engage in a role-play in which the helpee knows that he or she does not want help and the helper is trying to establish a helping relationship. (It might be helpful for the helpee to keep in mind something that he or she would not want to share under any circumstances, in order to play the role of reluctant client well.)

After each of you has played both roles, process your experiences, feelings, and reactions. In this exercise, most students learn that there is no way someone can make helpees reveal what they do not choose to reveal.

Exercise 9.9   What are the issues in the following cases? What would you do as the helper in each case? Role play these situations and discuss them in small groups.

1. You are a youth street worker, and you overhear a gang of boys talk about a burglary they committed yesterday. Later, one of the boys tells you that this gang is having trouble fencing some of the loot and asks you to help. He offers you a free color TV. It's taken you a long time to gain the trust of this boy.

2. You are working as a volunteer in a crafts program for the aged. An 85-year-old man with whom you've established a helping relationship tells you that he wants to marry a 78-year-old woman in the program. His children are raising strenuous objections and want to pull him out of the program to keep him away from "foolish temptation." He wants your help.

3. You are a white paraprofessional working in a community school as a vocational counseling aide. A black couple comes in with their 11th-grade son. They have just received notice from the guidance department that on the basis of an IQ test, their son is being sent to trade school for the rest of the year. The parents are very upset. They want him in a college-bound track.

4. You are a personnel worker, and Ms. Jones, a very competent administrative assistant in the advertising department, comes in to complain that a male administrative assistant with less background and experience has been promoted over her.

5. You are a helper in a drop-in center for teens. A 14-year-old girl tells you she thinks she is pregnant and that she was seduced by her employer. She says that her parents "will kill her" if they find out.

6. You are a male supervisor in a factory, supervising 14 male forklift operators. A woman is hired because of your factory's affirmative action plan. Your men are angry and upset and come to see you to plan how to stop this invasion of their territory.

7. You are a black helper in a youth activities program. A 15-year-old black youth who is bussed to a suburban school tells you he wants to ask a white girl to the school dance, but is somewhat nervous about it.

8. You are a school counselor in an upper-middle-class community. The parents of one of the students have come to see you and are very angry with you for encouraging their son in his interest in shop. They had always planned for him to go to medical school. The boy is a C student and is very talented in shop work.

9. You are a helper in a community center. Drugs are not allowed on the premises, but you find a youngster with whom you have been working smoking a joint.

10. You are a black counselor in a prison. You notice that only black prisoners are placed in behavior modification programs and that selection is arbitrarily determined by the prison administration. Subjects have no choice and are penalized if they do not cooperate.

☐ Common Problems You May Encounter as a Helper

Regardless of the amount of training and experience you have had, you will often feel insecure and doubt your adequacy as a helper. Perhaps this type of self-doubt is one of the necessary qualities of an effective helper; it keeps us on our toes, prevents us from becoming cocky and overconfident, and constantly reminds us of our own frailty as human beings. Working with people in a helping relationship is awesome and often frightening in that we become aware of the importance and vulnerability of human life. However, self-doubt can protect us from getting in over our heads, from attempting to work with people and problems beyond the scope of our training and capabilities.

Another problem that beginning helpers often experience is becoming too emotionally involved with helpees, wanting to take responsibility for them and do for them rather than teach them to do for themselves. It is because of this problem that we spend time developing our own self-awareness and examining our apparent and underlying motivations for becoming helpers as we study the helping process. As we learn about ourselves and about the helping process, we can learn to take care of ourselves and allow others to take care of themselves.

Helper **burnout** is an increasing problem phenomenon in the pressured human services field. When you feel exhausted and unable to pay attention to what someone is saying, or find yourself reacting more impatiently and intolerantly than you have in the past; when your sleep and eating habits change or when you experience a new physical symptom; when you find yourself dreading the beginning of the work day and lacking enthusiasm, motivation, and interest—you may be suffering from burnout.

Perhaps you will be able to make some changes in your work life to restore your energy and interest. For example, teachers can change the grade level or subject they teach, and school counselors can change their grade level. A clinical colleague of mine recently volunteered for major administrative and committee responsibilities to provide relief from full-time clinical practice. She intends to return to her clinical work but recognizes her need for temporary relief and distance from other people's emotional problems. Some helpers use burnout as an entry into their own personal therapy. Others undertake other stress-reduction activities, such as aerobic exercise, new hobbies or activities, or a vacation. Group work has been found to be effective, allowing helpers to share their concerns with each other or to associate with new, different people from other fields.

The point is to recognize beginning symptoms so that you can take steps to prevent the further effects of burnout. We need to recognize our own limitations, to learn how to say no to others and feel O.K. about it. We need to reassess our expectations in accordance with the changing realities of our work. In other words, like helpees, we need to take care of ourselves, to make time for our own inner selves so as to replenish and renew our energies and enthusiasm for living.

An additional problem you may encounter is that of total frustration with the limitations your organization or some of your community institutions impose on you and others. Learning what your options are and deciding what you want to do about them, what risks you can incur, and what resources you can muster can be painful and difficult and cause you to feel lonely and isolated from others. Along the same line, you will often feel anger and frustration with other members of the human services professions. You will question the quality and methods of some of the services provided. This kind of skepticism can occur in any area of endeavor associated with human beings and their welfare; eliminating it requires using your communication skills and knowledge to consult effectively with other human services workers to see if you can bring about change. The problem-solving model is especially applicable here in that it can show you how to team up with others to identify problems and brainstorm for alternative solutions.

A final problem I want to mention is your own resistance to changes and new ways of delivering services. People sometimes get so used to doing things in a certain way that without realizing it they become very set in their ways. Keeping open channels of communication with peers in your setting and others, attending meetings and conferences, reading new materials, and taking advantage of in-service training whenever possible will help you keep abreast of new developments and opportunities in your field.

Exercise 9.10   Take a few minutes to write yourself a personal contract for the coming week. What will you do to be good to yourself, to be your own helper? Try to contract for at least one way of reducing stress each day, one way of taking time for your own inner self.

☐ Recent Trends

Over the years the evidence of the effectiveness of beginning professional, paraprofessional, and lay helpers has been growing. There is evidence that the helping skills and strategies we have discussed are applicable, to varying degrees, to all three groups of helpers.

For example, telephone crisis intervention services report that short telephone calls using the communication skills we've discussed can provide sufficient help to callers to avert a pending crisis, such as a suicide or a crime. Often trusting, understanding human contact alone is sufficient to begin the helping process. Also, a recently developed telephone career-education counseling project demonstrated that vocational counseling can be effectively delivered by trained lay helpers via telephone. The use of computer technology for career counseling is also increasing.

Likewise, in many school and mental health settings a team approach involving professional, paraprofessional, and lay helpers has proven to be an effective way to provide human services. Professional helpers have reported that lay and paraprofessional helpers are invaluable in developing helping relationships and effecting behavior and attitude changes among client populations.

Industry is gravitating toward developing its human resource staffs to provide helping services to employees. Many organizations now provide health counseling, outplacement and retirement counseling, career counseling, and alcoholism and drug-abuse counseling. Human resource staffs often consist of professional counselors or social workers who have knowledge of organizational behavior as well as counseling theory and application. Some companies contract with outside helpers for employee assistance programs. The private business sector is steadily increasing its use of professional helpers. Another trend is the use of helpers in traditional health settings, providing such services as nutritional or stress-reduction programs and other preventive health measures. Holistic health is another emerging approach using interdisciplinary teamwork.

However, in spite of improvements in the quality and availability of counseling services, the need remains to demystify the helping process and expand in-service and community-based training opportunities so that increasing numbers of human services and community workers can deliver services on their own "turf." Only by so doing do we have the slightest chance of being able to provide vital human services to everyone who needs them. Every community possesses indigenous helping resources that need to be identified and developed. At the same time, supervisory and consultant-training services should be offered to professional helpers so that they can provide supervisory services as well as more intensive care when necessary.

Deinstitutionalization of the elderly, the mentally ill, prisoners, retardates, and the physically handicapped, along with individual financial reverses due to unemployment and bankruptcy, have created a larger homeless population than heretofore. Immediately, we need to recruit volunteer helpers from within the community to assist their neighbors in accepting people who have traditionally been isolated. In addition to needing trained helpers to assist our own ill, impoverished, substance-abusing, and otherwise handicapped persons, we need more trained helpers of all disciplines and levels to help returning veterans and their families as well as refugees and immigrants from all over the world. Never before has the need for effective helpers been so critical to our survival as a society.

☐ Summary and Conclusions

In this chapter we have discussed some of the major personal values and professional issues that affect helping relationships. As helpers we must be aware of our values and be careful not to impose them on clients. We can use responsive listening and send "I" messages to communicate respect for helpees and their values. Three issues discussed at some length were sexism, racism, and ageism. There are no absolute answers to the questions raised by those issues, but we can keep informed and maintain heightened sensitivity by continuing to discuss them. This process is as important as learning communication skills and understanding the stages of the helping relationship.

This chapter also discussed personal, professional, and ethical considerations, and recent trends in counseling. How we resolve the problems of privileged communication, confidentiality and duty to warn, conflict of interests, dual relationships, misrepresentation, record keeping, and testing is one of the factors that determine our effectiveness as helpers. How we take care of ourselves so we are capable of continuing to provide ethical, competent help is another such factor.

Throughout this book, we have had the opportunity to apply the communication skills, our understanding of the steps in the helping relationship stages, and the strategies through written and verbal exercises for both individuals and groups. These exercises have provided the opportunity to personally experience and react to the material in the text. They were designed to increase our awareness of our behaviors, thoughts, and feelings as well as to give us a chance to integrate conceptual material with experience.

Whether or not the case material has been appropriate to your setting and level of helping, it illustrates the stages of helping and many of the issues discussed. Most of the examples and case studies are from my own, my trainees', and my colleagues'practices. However, this material does not, as a model, favor any particular approaches over others.

Now that you have a foundation, you can decide whether or not to further develop your skills and your understanding of the helping process and the use of strategies. If you wish to become a more effective helper, it is important for you to receive supervised field experience as well as pursue academic courses in helping and the social sciences. Only by continual practice and application, however, can you improve your communication skills. Reading about and studying them can only serve as a foundation; laboratory and fieldwork application are essential.

Strategy application can be learned in course work, fieldwork, and specialized training institutes and workshops. In the past few years, continuing education has become big business, due to its being required for professional licensure. Although the quality varies, carefully selected programs can offer excellent training opportunities for discerning helpers. These workshops are often sponsored by professional associations at local, regional, and national conferences. Some are sponsored by local human relations groups or by professional institutes. Keeping an eye on local announcements and human services bulletins will alert you to available opportunities.

Certain approaches require more advanced training and experience than others. However, this requirement does not preclude you from having a basic understanding of all the major approaches. Having this basic understanding can serve as a stimulus to pursuing further training in a particular area now or at any other time during your activity as a helper. As training and other opportunities open up, there is more chance for lay helpers to become paraprofessionals and for paraprofessionals to become professionals. One need only look at the ages and backgrounds of students enrolled in different levels of training programs to recognize that people are coming into human services fields at all ages, with all kinds of experiences and backgrounds.

☐ References and Further Reading

Alinsky, S. (1972) *Rules for radicals.* New York: Vintage Books.

American Association for Counseling and Development. (1988). *Ethical standards* (rev. ed.). Alexandria, VA: Author.

American Mental Health Counselors Association. (1980). *Code of ethics for certified clinical mental health counselors.* Falls Church, VA: Author.

American Psychological Association, Task Force on Sex Bias and Sex-Role Stereotyping in Psychotherapeutic Practice. (1978). Guidelines for therapy with women. *American Psychologist, 33,* 1122–1133.

American Psychological Association. (1989). *Ethical principles of psychologists* (rev. ed.). Washington, DC: Author.

American Rehabilitation Counseling Association. (1981) *Code of ethics.* Alexandria, VA: Author.

Axelson, J. A. (1985). *Counseling and development in a multicultural society.* Pacific Grove, CA: Brooks/Cole.

Ballou, M., & Gabalac, N. W. (1985). *A feminist position on mental health.* Springfield, IL: Charles C Thomas.

Bates, C. M., & Brodsky, A. M. (1989). *Sex in the therapy hour.* New York: Guilford Press.

Bernays, T., & Cantor, D. W. (Eds.) (1986). *The psychology of today's woman.* Hillsdale, NJ: The Analytic Press.

Broverman, I., Broverman, D., Clarkson, F., Rosenkrantz, P., & Vogel, S. (1970). Sex-role stereotypes and clinical judgments of mental health. *Journal of Consulting and Clinical Psychology, 34,* 1–7.

Corey, G., Corey, M. S., & Callanan, P. (1988). *Issues and ethics in the helping professions* (3rd ed.). Pacific Grove, CA: Brooks/Cole.

Farber, B. A. (1983). *Stress and burnout in the human service professions.* New York: Pergamon Press.

Gartrell, N., Herman, J., Olarte, S., Feldstein, M., & Localio, R. (1987). Reporting practices of psychiatrists who knew of sexual misconduct by colleagues. *American Journal of Orthopsychiatry, 57,* 126–131.

Gilligan, C. (1982) *In a different voice.* Cambridge, MA: Harvard University Press.

Glaser, B., & Kirschenbaum, H. (1980). Using values clarification in counseling settings. *Personnel and Guidance Journal, 58,* 569–576.

Heilbrun, A. (1970). Toward resolution of the dependency–premature termination paradox for females in psychotherapy. *Journal of Consulting and Clinical Psychology, 34,* 382–386.

Herr, E. L., & Niles, S. (1988). The values of counseling: Three domains. *Counseling and Values, 33,* 4–17.

Kilburg, R. R., Nathan, P. E., & Thoreson, R. W. (Eds.). (1986). *Professionals in distress: Issues, syndromes, and solutions in psychology.* Washington, DC: American Psychological Association.

Kitchener, K. S. (1988). Dual role relationships: What makes them so problematic? *Journal of Counseling and Development, 67,* 217–221.

Lichtenberg, J. W., & Heck, E. G. (1983). Sex bias in counseling: A reply and critique. *Personnel and Guidance Journal, 62,* 102–105.

Miller, J. B. (1976). *Toward a new psychology of women.* Boston: Beacon Press.

National Association of Social Workers. (1979). *Code of ethics.* Washington, DC: Author.

National Association of Social Workers. (1981). *Standards for the private practice of clinical social work.* Washington, DC: Author.

Offir, C. (July 1974). At 65 work becomes a four-letter word. *Psychology Today,* 40–42.

Okun, B. (1984). *Working with adults: Individual, family, and career development.* Pacific Grove, CA: Brooks/Cole.

Okun, B. F. (1989). Therapist blindspots related to gender socialization. In D. Kantor & B. F. Okun (Eds.), *Intimate environments: Sex, intimacy, and gender in families* (pp. 129–163). New York: Guilford Press.

O'Malley, K. M., & Richardson, S. (1985). Sex bias in counseling: Have things changed? *Journal of Counseling and Development, 63,* 294–300.

Parker, G. (1967). Some concomitants of therapist dominance in the psychotherapy interview. *Journal of Consulting Psychology, 31,* 313–318.

Pedersen, P. (1988). *A handbook for developing multicultural awareness.* Alexandria, VA: American Association for Counseling and Development.

Pope, K. S. (1988). How clients are harmed by sexual contact with mental health professionals: The syndrome and its prevalence. *Journal of Counseling and Development, 67,* 222–226.

Rogers, C. (1967). The necessary and sufficient conditions of therapeutic personality change. *Journal of Consulting Psychology, 21,* 95–103.

Rosewater, L. B., & Walker, L. E. A. (Eds.). (1985). *Handbook of feminist therapy: Women's issues in psychotherapy.* New York: Springer.

Rubin, L. B. (1985). *Just friends: The role of friendship in our lives.* New York: Harper.

Sheridan, K. (1982). Sex bias in therapy: Are counselors immune? *Personnel and Guidance Journal, 61,* 81–83.

Simon, S., Howe, L., & Kirschenbaum, H. (1972). *Values clarification.* New York: Hart.

Smith, M. L. (1980). Sex bias in counseling and psychotherapy. *Psychological Bulletin, 87,* 392–407.

Sue, D. W. (1981). *Counseling the culturally different: Theory and practice.* New York: Wiley.

Sue, S. (1988). Psychotherapeutic services for ethnic minorities: Two decades of research findings. *American Psychologist, 43,* 301–308.

West, C. (1984). When the doctor is a "lady": Power, status, and gender in physician-patient encounters. *Symbolic Interaction, 7,* 87–106.

Whitely, B. E. (1979). Sex roles and psychotherapy: A current appraisal. *Psychological Bulletin, 86,* 1309–1321.

Glossary

Abreaction A Freudian term for the situation in which a patient relives painful emotional experiences during therapy, becoming conscious of previously repressed material.

Affective Pertaining to feelings and emotions. The affective domain comprises the feeling, emotional aspects of experience. These feelings may be conscious or unconscious.

Anal stage In psychoanalysis, the developmental stage between 2 and 3 years of age in which the child focuses on pleasure from the anal erogenous zone. This is a pregenital phase of sexual development during which bowel training becomes important.

Anima According to Jung, the feminine side of men and women.

Animus According to Jung, the masculine side of women and men.

Anxiety A state of tension that warns us of impending danger. Anxiety may be realistic, involving fear of danger from the external world; neurotic, involving fear that one's instincts will get out of control and lead to a punishable act; or moral, involving fear of one's own conscience. According to Freudian theory, anxiety results from the repression of the basic conflicts among the id, the ego, and the superego.

Approach reaction Behavior directed toward a situation or stimulus, regardless of positive or negative emotions. This tendency to deal with whatever issues are at hand implies a positive ability to work through difficulties.

Assertiveness training A behavioral technique whereby the client learns progressively more assertive behaviors (standing up for one's own rights without impinging on the rights of others) through such means as modeling, role playing, and instruction. Assertive behaviors include saying no without feeling guilty and learning to ask directly for what one wants.

Avoidance reaction   Withdrawal from or avoidance of a situation or stimulus that might have threatening or adverse emotional aspects. This reaction indicates a refusal to work through or check out problems and issues.

Behavioral   Pertaining to an observable physical action or performance; a concrete response one makes to a stimulus situation. Behaviors may be motor, perceptual, or glandular.

Birth order   Adlerian concept that one's placement within family structure (that is, whether one is the first-born child, the second, and so on) is a major determinant of personality.

Brainstorming   A group problem-solving technique; every person's idea is considered before evaluative screening occurs. This is an important step in problem solving, allowing for all possible input before a decision is reached.

Burnout   Depletion of physical and mental resources resulting in loss of motivation, interest, and capabilities for helping relationships.

Castration complex   The fear of losing the penis. Psychoanalysts believe that boys experience this fear during the phallic stage after they discover that girls do not have penises. The idea of this deprivation causes fear and anxiety and may be partly or wholly repressed.

Circular causality   The idea that there is a reciprocal connection between cause and effect, a cyclical interaction. This idea is the basis of systems theory.

Clarification   A verbal response that helps the client to better understand issues and needs through what has been said or felt.

Client-centered   Rogerian term meaning that the direction of the counseling (goals, course, and process) is wholly determined by the client, not the counselor.

Cognitive   Pertaining to thinking and knowing. Covers all modes of knowing: perceiving, remembering, imagining, conceiving, judging, reasoning. Cognition is a conscious process.

Cognitive restructuring   Rational-emotive therapy technique that identifies irrational thinking and replaces it with rational thinking through didactic teaching. One is taught how to correct faulty belief systems by unlearning irrational beliefs and learning rational ones.

Compensation   In psychoanalysis, a defense mechanism whereby one substitutes a satisfying activity for a frustrating one to reduce tension. One covers up a weakness or defect by displaying in exaggerated fashion a less defective or more desirable characteristic.

Concrete reinforcement   A specific object, such as a desired piece of food, or a special event, such as watching television, used as a reward for exhibiting designated behavior.

Conditioned response   A response that is learned, as opposed to one that is instinctive. Conditioning is a process in which a response is elicited by a stimulus, object, or situation; in other words, it is not a natural reflex.

Confrontation   A verbal technique whereby helpers present helpees with discrepancies between their verbal and nonverbal behaviors or between the helpee's and the helper's perceptions. Confrontation is often used to encourage approach, rather than avoidance, reactions.

Congruence   A client-centered therapy concept indicating agreement between one's experience and one's perceptions of that experience, or one's self-awareness. In other words, one's behavior is in tune with one's values and beliefs. One is said to "practice what one preaches."

Constructivism   A cognitive-behavioral approach focusing on people's active creation of and attribution of meaning to their personal and social realities.

Contamination   In transactional analysis, the situation that occurs when one of the ego states (parent, adult, or child) overlaps another and interferes with the effective functioning of that other ego state. Racial prejudice is an example of contamination, in that the parent ego state is saying that all members of one race are bad, and the adult ego state does not check that out with the real world, but merely accepts what was learned during childhood.

Content   The material or constituents of an experience, as distinct from the form or process of an experience. The "what" as opposed to the "how."

Contingency contracting   A positive behavioral contract between helper and helpee specifically stating desired behavioral outcomes and consequential reinforcements to follow performance of each stated behavior. An elaboration of the "If you do X, you will get (to do) Y" formula that is clearly stated, agreed to by all parties, and systematically applied.

Countertransference   In psychoanalysis, the positive and/or negative distortion of the analyst's interpretations of the client by the analyst's own conflicts.

Defense mechanism   Psychoanalytic construct of unconscious or involuntary strategies that one uses to protect oneself against painful negative feelings associated with a situation that is extremely disagreeable. The situation may be physical or mental and may occur frequently or infrequently.

Denial   In psychoanalysis, a defense mechanism whereby one's mind refuses to acknowledge and experience something that would cause anxiety and distress if acknowledged. It manifests itself as a firm belief that what happened did not happen or is not so.

Determinism   In psychoanalysis, the assumption that every mental event or attitude was caused by earlier psychological experiences and biological factors.

Directive therapy   A form of therapy in which the therapist guides the course, objectives, and process of therapy. The therapist is a director, a teacher, and a guide who has full control of the therapeutic relationship.

Discrimination   Learning theory term involving the ability to differentiate among slightly different stimuli. Reinforcement will have been present or stronger for some stimuli than for others.

Displacement   In psychoanalysis, a defense mechanism whereby the psychic energy (often anger or some other emotion) directed toward a particular person or object is transferred to another similar person or object. Displacement frequently occurs in dreams, in which feelings are shifted from one object or person to another object or person to which they do not really apply.

Dissonance   What occurs when two parts or aspects of experience do not blend or fuse, resulting in discrepancies and discomfort. Dissonance may occur between two behaviors, between behavior and feeling, or between inner and outer experiences.

Ego   Psychoanalytic concept of the partly preconscious, partly conscious part of the personality that is in direct contact with the external world, with the reality of the world. It includes conscious perceptions of reality as given by the senses and preconscious memories, together with those selected impulses and influences from within that have been accepted and are under control.

Ego states   According to transactional analysis, the three parts of the personality: the parent (providing criticism and nurturance), the adult (providing thought and reason), and the child (providing emotion, intuition, and creativity). These ego states are conscious and observable, and one can learn to identify and choose the one appropriate for a particular situation.

Electra complex   Psychoanalytic concept that describes a girl's phallic-stage jealousy of her mother's sex role. This jealousy causes the girl to direct her erotic feelings toward her father. The feelings are acted out with the girl expressing attachment to her father and antagonism to her mother.

Empathy   The ability to see the world the way the helpee sees it, from the helpee's "frame of reference."

Empty seat   Gestalt technique in which a client uses an empty chair to represent an imagined partner in a dialogue or a role-play game. The client sits in the "empty seat" when speaking as the person the seat represents.

Exclusion   Transactional analysis term that describes the situation that occurs when one of the three ego states (parent, adult, child) is missing and not functioning. This results in imbalance of the personality system and ensuing faulty transaction and communication.

Existential   Philosophical viewpoint focusing on the here and now, the presence of time, people's freedom to choose for themselves, purposes of life, potential, humanness. Existential psychology deals with those aspects of experience that can be observed introspectively (sensory and imaginal processes) together with feelings.

Extinguish   Behavioral technique in which any reinforcing consequences of a particular behavior are discontinued, resulting in that behavior being diminished and eventually discontinued. Extinction techniques are used to unlearn (erase) a specific behavior that is deemed inappropriate.

Extroversion   Jungian term that describes the state of being oriented toward the external and objective world. An extrovert is one who is primarily interested in things outside the self.

Free association   Psychoanalytic technique in which clients tell analysts everything that comes into their minds in response to a word or concept stimulus.

Generalization   Learning theory term for a response learned in one situation that can be used in other situations with similar but different stimuli; also refers to a general concept formed on the basis of several component ideas.

Genital stage   In psychoanalysis, the adult psychosexual stage of development that occurs as a child reaches puberty and his or her interests become heterosexual rather than self-centered. At this stage, the earlier psychosexual stages are fused, and genital eroticism predominates.

Gestalt   German word meaning *configuration*—that is, the form, pattern, structure, or configuration of an integrated whole, not a mere summation of its parts. Gestalt psychology contends that mental processes and behavior cannot be analyzed into separate elementary units because the human personality is a unified structure, a whole. Gestalt psychology is a psychology of perception.

Id   Psychoanalytic concept of primitive, unconscious source of psychic energy, a part of the personality that demands immediate gratification in order to reduce tension and increase pleasure. The inner determinant of conscious life.

Imagery techniques   Techniques that help people to imagine other people, scenes, and events, to recall or imagine sights, smells, feelings, and thoughts as vividly as possible.

Implosive therapy   A form of behavioral therapy in which the client vividly imagines intense exposure to aversive stimuli in order to extinguish anxiety about those stimuli. After repeated imaginary exposures, the client is able to deal with the aversive stimuli in real life with reduced anxiety.

Incongruence   In client-centered psychology, a term that refers to a discrepancy between a person's actual experience and his or her version of the experience. If someone's behavior is incongruent, it is not in keeping with what he or she says.

Inherent inferiority   Term in Adlerian psychology that refers to the belief that everyone comes into this life with built-in feelings of inferiority (based on helplessness, smallness, dependency). This inferiority provides the ultimate driving force of humans in that they strive to overcome inferiority and achieve superiority.

Intellectualization   In psychoanalytic psychology, a defense mechanism whereby one emphasizes intellectuality or cognition and neglects emotion and volition to avoid experiencing and dealing with emotional content.

Interpretation   A psychotherapeutic technique that uncovers the meanings and relationships underlying the apparent verbal content of a helpee statement or behavior. This technique may involve connecting different aspects of experience for the helpee or explaining relationships and causal factors.

Introversion   Jungian term that describes the state of being oriented toward the internal and subjective world. An introvert is concerned primarily with his or her own thoughts and feelings.

Isolation   In psychoanalytic psychology, a defense mechanism used by people to detach an idea from its affective or emotional content; for example, isolation may be indicated by a blank pause between a highly unpleasant or personally significant experience and a person's reaction to it.

Latency stage   Psychoanalytic concept of the stage of psychosexual development, occurring between the age of 4 or 5 and emerging adolescence, that separates infantile from genital sexuality. During this period, there are no conscious sexual interests or activities.

Leading   A verbal skill that elicits client responses in an open-ended yet focused manner. Leads include "door openers" such as "Tell me more . . ." and "I'm wondering about . . ." as well as questions, reflections, clarifications, and so forth.

Libido   Psychoanalytic concept referring to psychic energy that includes the sexual and survival instincts. Libido is a dynamic force that includes both sexual and ego drives. It is the life force that serves to neutralize the destructive impulses in the system.

Lifestyle   In Adlerian psychology, a person's unique, directional pattern of behavior based on the process of judging both status of the self and status of the world.

Modeling   A learning theory principle whereby new modes of behavior are learned or old ones changed by observing others' actual or simulated behavior and its consequences. Modeling is based on imitation.

Neurosis   In the psychoanalytic sense, a functional disorder of the nervous system that is psychological rather than organic in origin. This disorder can result in somatic and behavioral symptoms. Because the adequate satisfaction of subjective needs is the function of the ego, neurosis can be understood to be a disturbance of ego functions.

Object   In psychoanalytic object relations, an object refers to one's mental representations of significant others who are sources of sustenance, protection, and gratification.

Oedipal complex   According to psychoanalytic theory, a disturbance that may occur during the phallic stage, when one directs his or her erotic desires toward the opposite-gender parent. One expresses antagonism to the same-gender parent and attachment to the opposite-gender parent.

Operant conditioning   A principle of behavioral learning theory whereby the actual consequences of behavior, rather than the original cause (stimulus), are of concern. These consequences of a behavior "operate" on the behavior and on the environment, causing the behavior to recur.

Oral stage   Psychoanalytic term for the infantile stage of psychosexual development when pleasure is centered around the mouth, around sucking and eating activities.

Organismic   Pertaining to the individual as a whole entity; emphasis is placed on the organized system of interrelated and interdependent parts.

Penis envy   Psychoanalytic concept describing little girls' envy of boys, and their desire to have a penis.

Permission   In transactional analysis, a term that describes the situation in which the helper's or helpee's parent ego state allows a helpee to feel, be, or act as he or she is or chooses. In most cases, the helper activates the helpee's nurturing parent to give permission to the helpee's child ego state to be whole and free, to feel, and so on.

Persona   The Jungian term for the disguise or mask that one displays in public and that frequently differs from one's true attitude or appearance.

Phallic stage   In psychoanalysis, a developmental stage in which a child's interests shift from the anal to the genital area. According to Freud, during this stage, children believe that both males and females have penises.

Phenomenological   Perceiving someone else's world through his or her eyes; emphasizing conscious experience as real experience.

Positive reinforcement   Rewards that have good significance (for example, money or good grades). Positive reinforcement may be social or concrete.

Potency   A term used in Gestalt psychology and transactional analysis that implies that the helper or helpee has the power to do something, has expertise and credibility to deliver.

Principle of gradation   Learning theory principle whereby a sequence of intermediate or subgoals leads gradually to a more complex behavior.

Process   A series of successive but interdependent changes or events. Also refers to how something is happening and what the associated effect and form are as opposed to the actual facts and events (processing).

Projection   In psychoanalytic theory, a defense mechanism whereby one unconsciously attributes feelings (for example, guilt or inferiority), thoughts, or acts unacceptable to one's own ego to other people. These projections are usually a defense against unpleasant feelings in ourselves and are means by which we can justify ourselves in our own eyes.

Projective techniques   Tests or instruments in the form of stories, pictures, or role playing that raise unconscious material to consciousness. This type of mental testing helps to determine personality traits. The individual is left free to follow his or her inclinations and fantasies.

Protection   In transactional analysis, the helper's ability to take care of helpees and keep them safe. Helpers activate helpees' nurturing parent to protect their child ego state so that helpees will not be hurt by their (own) actions.

Psychoeducation   Helping strategy that teaches people about areas of emotional and relationship functioning—that is, how people learn and develop.

Psychosexual stages   Psychoanalytic developmental stages that are crucial in the child's personality development. Because the Freudian view places sexual impulses at the root of all human personality problems, sexual development and sexually oriented experiences in childhood directly affect future development.

Racket   In transactional analysis, a collection of feelings that one uses to justify one's life "script" and one's decisions; one uses rackets to set up situations that give rise to the collected feelings as a "payoff." Rackets are learned in childhood and persist into adulthood.

Rationalization   In psychoanalysis, a defense mechanism whereby a person assigns a socially acceptable motive to his or her behavior. The process of rationally justifying an act helps one defend oneself against self-accusation or guilt.

Reaction-formation   A psychoanalytic term for behavior that is directly opposed to unconscious wishes; this behavior represents a defense mechanism that allows the ego to keep unacceptable character traits in check. Reaction-formation behavior can sometimes be excessive or violent.

Recall   A psychoanalytic term meaning to revive or reinstate a past experience in memory.

Referral   The arrangement of other assistance for a helpee when the initial helping situation is not or cannot be effective.

Reframing   Technique that relabels behavior in a more positive framework—for example, "When you fight, you're expressing caring." This changes people's perspectives and allows new responses.

Regression   Psychoanalytic term that describes the retreat to an earlier stage of development in which one felt more adequate; one expresses interests and behavior characteristic of an earlier stage. Regression is a defense mechanism used to reduce tension or anxiety.

Reinforcement   Behavioral term for the environmental event that, when following certain behavior, causes that behavior to recur. May be positive or negative. Reinforcements are the environmental consequences of behavior.

Repression   In psychoanalysis, a defense mechanism whereby one forces into the unconscious painful perceptions, ideas, and feelings. Impulses and desires in conflict with enforced standards of conduct are thrust into the unconscious where they can still remain active, indirectly determining behavior and experience, perhaps through dreams or neurotic symptoms. According to Freud, repression is often the cause of neurotic disorders.

Resistance   Defensive behavior of the helpee that prevents him or her from participating effectively in helping relationships and process.

Rubberband   Transactional analysis concept of a feeling associated with an early event in one's life that one feels strongly or that causes one to overreact in response to a present situation.

Schedules of reinforcement   Behavioral term describing reinforcement schedules used in contingency contracting: (1) continual reinforcement—reinforcement that follows each time target behavior occurs; (2) interval schedule–reinforcement that occurs after a certain period of time, such as every hour; and (3) ratio schedule—reinforcement that occurs after a certain number of responses, such as on every third response.

Script   Transactional analysis concept of early life decisions that result in lifelong patterns and themes reflected in behaviors, feelings, and thoughts; a life "plan" determined early in life. Every individual has many scripts—to be a success or to be a failure, to be perfect—that dominate his or her life until they are brought into awareness by rescripting early childhood decisions.

Self-concept   According to client-centered theory, the perception we have of ourselves based on information from significant others and from our experiences. Our image of who and what we are, what we are all about.

Shaping   Learning theory term for the modification of behavior by reinforcing more and more refined responses that come closer and closer to the desired behavior. One can shape behavior by breaking it into its smallest parts and reinforcing one part at a time, in sequence.

Significant other   A parent, relative, teacher, or other person who is especially meaningful and important to an individual. Significant others have influence over our feelings and actions.

Social reinforcement   Behavioral term meaning attention from a significant other (for example, a smile, nod, physical contact, praise) that follows a particular behavior and makes that behavior more likely to recur.

Splitting   Psychoanalytic object relations term for separating off a part of the self (or ego) as a polar opposite of the rest, or being oblivious to one part of the self (or ego).

Stroke   Transactional analysis concept of reinforcement; any unit of recognition a person receives from another person. Strokes may be positive or negative, conditional or unconditional.

Sublimation   According to psychoanalytic psychology, a defense mechanism that gives blocked energy an alternative, socially acceptable outlet; for example, instead of seeking sexual gratification, one might help people or engage in artistic endeavors.

Successive approximation   A learning theory concept whereby a desired behavior is broken into its smallest parts, and behaviors resembling one of those parts are reinforced. For example, the act of holding a pencil in the writing hand may be the first behavioral component of learning to write. This or other resembling behaviors can be reinforced in order to shape the desired target behavior.

Superego   A psychoanalytic term for the part of the personality that interjects parental and social values. The superego is a structure in the unconscious built up by early experiences with parents and significant others. When the ego gratifies the primitive impulses of the id, the superego criticizes the ego, which results in feelings of guilt and anxiety.

Sweatshirt   Transactional analysis concept meaning the messages one broadcasts about oneself by manner and behavior, usually without knowing it. Sweatshirts usually have two messages, one on the front and one on the back, such as "Come on and get me" on the front and "How dare you?" on the back.

Systematic desensitization   Behavioral technique of counterconditioning to reduce anxiety by associating negative stimuli with positive experiences so that the stimuli no longer arouse anxiety. Desensitization begins when a client learns complete muscle relaxation (which is antithetical to anxiety) and establishes an anxiety hierarchy. An anxiety-causing stimulus is then paired with positive mental images and the process of relaxation. The pairing continues until the entire hierarchy can be imagined without anxiety.

Token economy   A behavioral reinforcement program in which desired behaviors are reinforced with tokens that can be exchanged for rewards. The tokens can be awarded immediately after the behavior is performed, and the exchange of tokens can occur later.

Transference   Psychoanalytic term for the situation in which the client unconsciously puts the analyst in the place of one or more significant others in his or her life and attributes to the analyst the attitudes, behaviors, and attributes of the significant person(s). Transference refers to the patient's developing a positive or negative emotional attitude toward the analyst.

Unfinished business   Gestalt term for unexpressed feelings, associated with memories or fantasies, that affect current functioning.

Appendix A:
Observer's
Guide to Rating
Communication Skills

This observer's guide can be used in conjunction with the exercises on communication skills in Chapters 3 and 4. The helper will not use all the behaviors listed for every role-play, but over a period of time the ratings will indicate verbal and nonverbal behaviors that need further development as well as those that the helper is already using effectively. This guide will help both observer and helper understand and recognize the communication behaviors that are necessary to make helping relationships effective.

Rate the helper's behaviors on a scale of 0 to 3: 0 = did not occur, 1 = occurred but needs improvement, 2 = occurred and is adequate, 3 = helper especially strong on this point.

☐ Nonverbal Behaviors

1. The helper maintained eye contact with the helpee.
   0   1   2   3
2. The helper varied facial expressions during the interview.
   0   1   2   3
3. The helper responded to the helpee with alertness and facial animation.
   0   1   2   3
4. The helper sometimes nodded his or her head.
   0   1   2   3
5. The helper had a relaxed body position.
   0   1   2   3
6. The helper leaned toward the helpee to encourage the helpee.
   0   1   2   3
7. The helper's voice pitch varied when talking.
   0   1   2   3

8. The helper's voice was easily heard by the helpee.

   0   1   2   3

9. Sometimes the helper used one-word comments such as "mm-mm" or "uh-huh" to encourage the helpee.

   0   1   2   3

10. The helper communicated warmth, concern, and empathy by smiling and using other gestures.

    0   1   2   3

☐ Verbal Behaviors

11. The helper responded to the most important theme of each of the helpee's statements.

    0   1   2   3

12. The helper usually identified and responded to the feelings of the helpee.

    0   1   2   3

13. The helper usually identified and responded to the behaviors of the helpee.

    0   1   2   3

14. The helper verbally responded to at least one nonverbal cue from the helpee.

    0   1   2   3

15. The helper encouraged the helpee to talk about his or her feelings.

    0   1   2   3

16. The helper asked questions that could not be answered in a yes-or-no fashion.

    0   1   2   3

17. The helper confronted the helpee with any discrepancies between behavior and communication.

    0   1   2   3

18. The helper shared his or her feelings with the helpee.

    0   1   2   3

19. The helper communicated understanding of the helpee.

    0   1   2   3

20. The helper responded in ways that communicated liking for and appreciation of the helpee.

    0   1   2   3

21. The helper summarized statements and themes to clarify issues for the helpee.

    0   1   2   3

22. The helper sent "I" messages when confronting the helpee or expressing lack of understanding.

    0   1   2   3

☐ Your Summary of Suggestions for Helper

Appendix B:
Sample
Psychotherapy Policies*

Please read this information carefully. It describes my policies and the related practices to be followed as part of the therapy services I will provide you. Be sure to raise any questions you may have. I would appreciate your returning a signed copy the next time we meet.

☐ Appointments and Cancellations

Every effort will be made to schedule appointments that are mutually convenient. If it becomes necessary for you to cancel, at least 24 hours notice must be given. If less than 24 hours notice is given, you will be expected to pay for that appointment. (Insurance companies will not reimburse for canceled or missed sessions.) There is no need to confirm appointments unless you have a question.

☐ Payment and Insurance

Please know your insurance coverage. Pay special attention to annual limits, deductive amounts you must pay yourself, percentage of charges your insurance pays, and any waiting period if your insurance is new.

While I will complete your insurer's claim forms, it remains your responsibility to guarantee payment and to follow up with your insurance company if there are any questions. Billing is generally by the month. It is expected that payments will be made on a timely basis—within one month of billing.

---

*This form is based on a model developed by Judith Birnbaum, Ph.D., Wellesley, MA.

You should be aware that for claims to be processed, insurance companies require a diagnosis and, occasionally, other information. By law, such information cannot be released by insurance companies without your specific, informed consent.

☐ Confidentiality

Communications between my clients and me are confidential, in accord with professional ethics and in compliance with the law. However, Massachusetts law also specifies certain limitations to this confidentiality. While these limits may not be at all relevant to your particular situation, I am legally obligated to inform you about them. The following are conditions in which disclosure can be made without your consent. I must disclose information:

1. In order to protect you or others if
   a. you present a danger to yourself and refuse to accept appropriate treatment;
   b. you tell me of an actual threat to harm another person;
   c. you have a history of violence and there is cause to believe you pose a danger of physical violence to another.
2. In case of child or elder abuse, which must be reported to appropriate state agencies.
3. In order to collect debts or to protect myself in a court action.
4. In certain legal proceedings should a court of law issue an order requiring the release of confidential information.
5. With colleagues about my work with you (never revealing your identity) to provide the best services possible. In any case, only appropriate and necessary information will be provided.

Of course, whenever you wish to give expressed, written consent, I can share information about you. When I am working with you and your family (or spouse), it is important that nothing anyone says during a session be used against him or her outside of the session. Likewise, I will never discuss one member of a family with another during any individual sessions that might occur adjunctively with couples or family work unless one of the above conditions is present.

(*Signed*) _____

(*Signed*) _____

(*Date*)   _____

Index

Abreaction, 110
Abrego, P. J., 32
Adler, Alfred, 104, 108, 111
Advocacy, 248, 254
Affective domain, 8, 141
Affective messages, 55–58
Affective strategies, 142–148
Affective-cognitive strategies, 148–149
Ageism, 247–248
American Association for Counseling
    and Development, 249
American Psychological Association, 249
Anal stage, 106
Anima, 107
Animus, 107
Approach reactions, 47
Assertiveness training, 162–163, 167–168
Authier, J., 10
Avoidance reactions, 41, 47, 116

BASIC ID model, 128, 170, 174–175, 179,
    218
Beck, A. T., 123, 152–153, 158, 159
Behavioral approaches, 8–9, 42,
    117–119, 153, 160–169
Behavioral domain, 8, 141
Berenson, B., 31–32
Birth order, 108
Brabeck, M., 126
Brainstorming, 150
Brammer, L., 32
Brief therapy, 220–221
Broverman, I., 241

Bruce, P., 138
Bugental, James, 111
Burnout, 257–258

Caplan, Gerald, 216, 217
Carkhuff, R., 10, 31–32
Case studies, 199–211
Cashdan, S., 148
Change-agentry, 254–255
Checking out, 70–71
Circular causality, 129
Clarification stage, 85–87
Clarifying, 70, 113
Client-centered approach, 8, 10, 11, 42,
    112–114, 142
Cognitive domain, 8, 141
Cognitive messages, 52–55
Cognitive restructuring, 9, 158
Cognitive strategies, 150–152
Cognitive-behavioral approaches,
    119–124, 152–159, 217
Cognitive-behavioral therapy (Beck),
    123–124, 152–153, 158, 159
Combs, A., 31
Communication, 15–16, 47–77
    in helping relationship, 23–30, 38
    importance of, 3–4, 5, 10
    nonverbal, 47–51, 59–60, 116
    responding, 15–16, 59–73
    and self-awareness, 41
    verbal, 51–58, 60–73
Confidentiality, 249–250
Conflicts of interest, 250–251

Confronting, 70, 71, 113, 226
Congruence, 11, 37–38
Constructivism, 119
Content, 12, 43, 51
Contingency contracts, 118
Contracting, 87–90, 161–162, 167
Control, 41
Corey, G., 30, 32, 39
Counseling, definition of, 10
Countertransference, 110–111
Crisis clinics, 222
Crisis intervention, 212–229, 258–259
    prevention, 224–225
    theory, 216–218
Cross-cultural models, 127. See also
    Diversity issues

DCT. See Developmental counseling and
    therapy
Defense mechanisms, 105
Definition of structure/contract stage,
    87–90
Deinstitutionalization, 259
Determinism, 105
Developmental counseling and therapy
    (DCT), 128–129, 175
Dialogue, 13, 144–145, 147
Discrimination, 118
Dissonance, 11
Diversity issues, 1, 2, 38–39
    ageism, 247–248
    and developmental counseling model,
        33
    and empathy, 7–8
    helper sensitivity, 5, 35–36
    and nonverbal communication, 51
    racism and ethnocentrism, 244–247
    sexism, 240–244
Domains, 8, 138–142
Doyle, R. E., 72
Drop-in centers, 222
Dual role relationships, 251–252
Dworkin, S., 127

Eclectic strategies, 169–175
Egan, G., 10, 32
Ego, 105
Ego psychology, 108–109

Ego states, 124
Eichenbaum, L., 127
Electra complex, 106
Ellis, Albert, 120, 152, 153, 154
Empathy, 7–8, 111
Empty seat, 143, 148
Erikson, Erik, 104, 108
Ethical integrity, 39–40
Ethics, 249–253
Ethnic differences. See Diversity issues
Ethnocentrism, 244–247. See also
    Diversity issues
Evaluation, 190–192
Exercises
    affective-cognitive strategies, 149
    behavioral strategies, 166–169, 180
    burnout, 258
    choosing strategies, 175–179, 180–181
    cognitive strategies, 152
    cognitive-behavioral strategies, 159
    crisis intervention, 213, 214–215, 216,
        218, 224–225, 228–229
    Gestalt strategies, 147–148
    helping relationship, 24–30, 43–45,
        84, 94–100, 101
    issues, 255–257
    nonverbal communication, 48–51,
        59–60, 74
    personal values, 237–240, 243–244
    strategy stages, 187, 189, 190, 192,
        195–196, 196–199
    systems strategies, 174–175, 180
    TA strategies, 172
    verbal communication, 52–55, 56–58,
        61–73, 74–77
Existential influences, 11
Existential theory, 112
Extroversion, 107

Fairbairn, Ronald, 109
Fantasizing, 147
Feldstein, M., 242
Feminist movement, 1
Feminist therapies, 126–127
Fisch, R., 220
Follow-up, 196
Frankl, Viktor, 111
Freud, Anna, 108

Freud, Sigmund, 104, 105–107
Fromm, Erich, 104

Gartrell, N., 242
Gender. *See* Diversity issues; Feminist
    therapies
Generalization, 118
Genital stage, 106
Gestalt approach, 8, 13, 33, 111, 114–117,
    142
  nonverbal communication, 59–60
  strategies, 142–146
  See also Exercises
Gilligan, C., 127, 240
Glaser, B., 233
Glasser, William, 121, 122, 152
Goal establishment stage, 91–93, 185–187
Gordon, T., 10
Gradation, principle of, 117
Guntrip, Harry, 109

Hare-Mustin, R., 127
Hartmann, Heinz, 108
Heck, E. G., 241
Heilbrun, A., 241
Helper
  characteristics of, 30–40
  common problems, 257–258
  definitions, 5–7
  self-awareness, 5, 34–35, 40–41, 114
  See also Helping relationship;
    Strategies
Helping relationship, 12–13, 14–15,
    20–42, 78–100
  in behavioral theory, 118–119
  clarification stage, 85–87
  in cognitive-behavioral theory, 121,
    122–123, 124, 156–157
  communications in, 23–30
  conditions affecting, 78–82
  in crisis intervention, 219
  definition of structure/contract stage,
    87–90
  dual role, 251–252
  gender issues in, 241–242
  goal establishment stage, 91–93
  and helper characteristics, 30–40
  initiation/entry stage, 82–85

Helping relationship *(continued)*
  intensive exploration stage, 90–91
  kinds of, 21–22
  in phenomenological theory, 114, 116
  in psychodynamic theory, 111
  racism in, 245–246
  in TA theory, 125–126
  See also Communication; Strategies
Helping theories, 38–39, 102–130, 131,
    132
  behavioral, 8–9, 42, 117–119
  cognitive-behavioral, 9, 119–124
  integrative, 126–129
  phenomenological, 31, 111–117
  psychodynamic, 42, 104–111
  transactional analysis, 124–126
Herman, J., 242
Hidden agendas, 51
Homme, L., 161
Honesty, 36–37, 41
Horney, Karen, 104
Hot lines, 222–223, 258–259
Howe, L., 233
Human relations counseling model,
    11–17
Hummel, Flora, 163
Hutchins, D. E., 39

Id, 105
Imagery techniques, 121
Implosive therapy, 118
Incongruencies, 11
Informing, 71
Inherent inferiority, 108
Initiation/entry stage, 82–85
Integrative theories, 126–129
Intensive exploration stage, 90–91
Interpersonal problems, and
    communications, 23
Interpreting, 70, 71
Interviews, 78–79
Introversion, 107
Issues, 16–17, 232–259
  ethics, 249–253
  personal values, 232–249
  See also Diversity issues
Ivey, Allen E., 10, 32, 33, 35, 126, 128, 175
Ivey, M., 10, 32

Jung, Carl, 104
Jungian theory, 107–108

Kagan, N., 10, 67
Kernberg, Otto, 109
Kirschenbaum, H., 233
Kitchener, K. S., 249
Klein, Melanie, 109
Knowledge, 38–39
Koffka, Kurt, 114
Köhler, Wolfgang, 114
Kohut, Heinz, 109
Kottler, J. A., 30, 39
Krantzler, M., 226
Kübler-Ross, E., 226

Larrabee, M. J., 84
Latency stage, 106
Lazarus, Arnold, 127, 128, 218
Leading, 61
Levitsky, A., 143
Libido, 105
Lichtenberg, J. W., 241
Lifestyle, 108
Lindemann, Eric, 216, 217
Localio, R., 242

Mahler, Margaret, 109, 126
Managed health care, 2
May, Rollo, 111
Miller, J. B., 127, 240
Minimal verbal response, 70, 113
Misrepresentation, 252
Modeling, 61, 160–161, 167
Multimodal therapy, 127–128, 170,
    174–175, 179, 218

National Association of Social Workers,
    249
Neuroses, 105–106, 107
Nonprofessional helpers, 5, 7, 10, 21,
    258–259
Nonverbal communication, 15, 47–51,
    59–60, 116

Object relations, 109–110
Oedipal complex, 106
Okun, B. F., 30, 126, 127, 130, 148, 248

Olarte, S., 242
O'Malley, K. M., 241
Operant conditioning, 118
Oral stage, 106
Orbach, S., 127
Organismic balance, 115
Otani, A., 93
Outreach counseling, 223–224

Paraphrasing, 70, 113
Paraprofessional helpers, 5, 7, 10, 21,
    258–259
Parker, G., 241
Pedersen, P. A., 127, 246
Penis envy, 106
Perls, Fritz, 111, 114, 143
Permission, 125
Persona, 107
Personal values, 232–249
Phallic stage, 106
Phenomenological theory, 31, 111–117,
    142, 217
Potency, 116
Probing, 70
Problem ownership, 86, 116
Process, vs. content, 12
Processing, 47
Professional helpers, 5, 6, 10, 21, 258–259
Protection, 125
Psychodynamic theory, 42, 104–111, 148,
    217
Psychoeducation, 173
Psychosexual stages, 106
Psychosocial stress, 1–3

Racism, 244–247
Rappaport, L., 130
Rational-emotive therapy, 9, 120–121,
    152, 155–156, 158–159
Reality therapy, 121–123, 152, 156–158
Record keeping, 251
Referrals, 87–88
Reflecting, 70, 113
Reframing, 173, 174
Reich, Wilhelm, 104
Reinforcement, 11, 117, 167
Relationship. See Helping relationship
Relaxation, 163–165, 169

Reluctant clients, 23, 84–85, 253, 255–256
Reparenting, 171
Resistance, 93–94
Responsibility, 5, 20
Responsive listening, 15–16, 59–73, 146
Richardson, S., 241
Rogers, Carl, 10, 31, 111, 112–114, 142, 237
Rubberbands, 124
Rubin, L. B., 242

Scripts, 124
Self psychology, 109–110
Self-acceptance, 5
Self-awareness, 5, 34–35, 40–41, 114, 115, 142
Self-concept, 112–113
Self-disclosure, 113–114, 242, 253–254
Sexism, 240–244
Sexual harassment, 242–243
Sheridan, K., 241
Shostrum, E., 32
Shuttling, 147
Significant others, 11
Simek-Downing, L., 10, 32
Simon, S., 233
Smith, M. L., 241
Social networks, 218
Splitting, 109–110
Strategies, 14, 15, 138–179
    affective, 142–148
    affective-cognitive, 148–149
    applying, 185–211
    behavioral, 160–169
    case studies, 199–211
    cognitive, 150–152
    cognitive-behavioral, 152–159
    in crisis intervention, 220
    eclectic, 169–175
Strokes, 9, 124, 172
Successive approximations, 162
Sue, D. W., 127, 246

Sue, S., 127, 246
Sullivan, Harry Stack, 104
Summarizing, 71, 113
Superego, 105
Sweatshirts, 124
Systematic desensitization, 8, 163, 165–166
Systems perspective, 12, 20, 129–130, 173–175

TA. See Transactional analysis theory
Termination, 10, 15, 192–196
Testing, 251
Theories. See Helping theories
Token economy, 118
Training, 30–31
Transactional analysis (TA) theory, 124–126, 170–172
Transference, 110, 148–149
Trust, 7. See also Helping relationship

Unconditional positive regard, 113
Unfinished business, 115

Values clarification, 233–234
Values sharing, 234–236
Verbal communication, 15, 51–58, 60–73

Ward, D. E., 15
Ward, D. W., 194
Watkins, C. H., Jr., 148
Watzlawick, P., 220
Weakland, J., 220
Welfel, E., 126
Wertheimer, Max, 114
West, C., 242
Whitely, B. E., 241
Winnicott, D. W., 109
Wolpe, Joseph, 8

Yalom, Irving, 111

TO THE OWNER OF THIS BOOK:

I hope you have found *Effective Helping* (fourth edition) useful. So that this book can be improved in a future edition, would you take the time to complete this sheet and return it? Thanks.

School and address: _____

Department: _____ Instructor's name: _____

1. What I like *most* about this book is: _____

_____

_____

_____

2. What I like *least* about this book is: _____

_____

_____

_____

3. My general reaction to the cases in the book is: _____

_____

_____

4. Exercises I found most helpful are: _____

_____

5. Specific exercises I found least useful or would suggest deleting in future editions are:

_____

_____

6. My general reaction to this book is: _____

_____

7. The name of the course in which I used this book is: _____

8. On a separate sheet of paper, please write specific suggestions for improving this book and anything else you'd care to share about your experience in using the book.

Optional:

Your name: _____ Date: _____

May Brooks/Cole quote you, either in promotion for *Effective Helping* or in future publishing ventures?

      Yes: _____ No: _____

      Sincerely,
      Barbara F. Okun

FOLD HERE

FOLD HERE